CONFORMAL MAPPING

L. BIEBERBACH
Friedrich-Wilhelm's University

TRANSLATED BY
F. STEINHARDT
Columbia University

AMS CHELSEA PUBLISHING
American Mathematical Society • Providence, Rhode Island

2010 *Mathematics Subject Classification.* Primary 30-02, 30Cxx;
Secondary 01A75.

Library of Congress Cataloging-in-Publication Data

Bieberbach, Ludwig, 1886–
Einführung in die konforme Abbildung. English
Conformal mapping / L. Bieberbach ; translated by F. Steinhardt.
Originally published: New York : Chelsea Pub., 1953
Includes bibliographical references and index.
ISBN 0-8218-2105-9 (alk. paper)
1. Conformal mapping. I. Title.

QA646.B513 2000
516.3′6

Reprinted by the American Mathematical Society, 2000, 2015.
♾ The paper used in this book is acid-free and falls within the guidelines
established to ensure permanence and durability.
Visit the AMS home page at http://www.ams.org/

10 9 8 7 6 5 4 3 2 20 19 18 17 16 15

TRANSLATOR'S PREFACE

This book is a translation of the fourth (latest) edition of Bieberbach's well-known *Einführung in die Konforme Abbildung*, Berlin 1949. Various minor corrections have been made (particularly in §§ 12 and 14) and the bibliography has been added to.

F. Steinhardt

TABLE OF CONTENTS

I. Foundations. Linear Functions

II. Rational Functions

III. General Considerations

IV. Further Study of Mappings Represented by Given Formulas

V. MAPPINGS OF GIVEN REGIONS

CHAPTER ONE

Foundations. Linear Functions

§ 1. Analytic Functions and Conformal Mapping

As is well known, a function $w = f(z)$ of a complex variable $z = x + iy$ $(i = \sqrt{-1})$ is said to be *analytic* and *regular throughout a region*[1] R if it is one-valued and differentiable at every point of R. A consequence of the differentiability of $f(z) = u(x, y) + iv(x, y)$ are the Cauchy-Riemann differential equations for the real and imaginary parts of $f(z)$, viz.

$$\frac{\partial u}{\partial x} = \frac{\partial v}{\partial y}, \quad \frac{\partial u}{\partial y} = -\frac{\partial v}{\partial x}. \tag{1}$$

We further assume the reader to be familiar with the fact that analytic functions can be developed in power series, that is, that in the neighborhood

---

[1] A *region* is a point-set with the following two properties: 1. If a point P belongs to the set, then so do all the points of some circular disc that contains P in its interior. 2. Any two points of the set can be connected by a continuous curve all of whose points belong to the set.—A closed region, i.e. a region plus its boundary points, is sometimes called a *domain*; many writers, however, use the term "domain" to mean the same as "region."

of any given point a of the region R, an expansion of the form

$$(2) \qquad w = c_0 + c_1(z-a) + c_2(z-a)^2 + \cdots$$

holds. Now consider, in particular, functions $f(z)$ for which $f'(z) \neq 0$ holds everywhere in R, and interpret x, y and u, v as rectangular coordinates, in the usual way; it is proved in Function Theory that if R is mapped by $w = f(z)$ on a point-set R', then R' is itself a region (*Theorem on the Preservation of Neighborhoods*).

That is to say, 1. if c_0 denotes a point of R' which is such that the a in $c_0 = f(a)$ is an *interior* point of R, then all the points within a sufficiently small circle with center at c_0 also belong to the point-set R'; 2. the point-set R' is connected.

The first part of this theorem is merely the geometric expression of the fact that power series (with $c_1 \neq 0$) have an inverse. For, from $w = c_0 + c_1(z-a) + \ldots$ it follows that $z = a + (1/c_1)(w - c_0) + \ldots$. Let a be an interior point of R; then the last power series converges within some circle with center at $w = c_0$. By confining ourselves to a sufficiently small such circle, we can make sure that its points correspond under $w = f(z)$ to z-values from a neighborhood of $z = a$ that belongs to R. But then, the circle we have chosen must have all its points in R'.

The second part of our theorem states that any

two interior points of R' can be connected by a continuous curve consisting entirely of interior points of R'; but this follows immediately from the possibility of doing the same for the corresponding points of R and from the fact that $u(x, y)$ and $v(x, y)$ are continuous functions.

Remarks. 1. If the mapping function is regular on the boundary of the region as well, then the boundary points of R are mapped onto boundary points of R', since we would otherwise be led to a contradiction with the neighborhood-preserving character of the inverse mapping (of R' on R).

2. We have proved our theorem only under certain restrictions. We shall soon see that it holds for all functions that are regular except for poles, and that it also holds for infinite regions provided only that we extend our definition of region a little; see § 7.

3. It may happen that one and the same point of the w-plane occurs both as an interior point and as a boundary point of R'; this has to do with the possible many-valuedness of the inverse of the mapping function $f(z)$. It is not at all a foregone conclusion that $f(z)$ will assume every one of its values only once in R, and thus it may also happen that it assumes one and the same value in the interior and on the boundary of R. The point of the w-plane corresponding to such a value is then an interior point as well as a boundary point of R'.

This may at first make some trouble for one's visualization of the geometric situation. But ever since Riemann's time, this stumbling block to an intuitive geometric grasp has been overcome satisfactorily. What is needed here can be made clear as follows. In Fig. 1, imagine a long "tongue" attached to the rectangle along AB, and let that tongue overlap the rectangle in the shaded part. This constitutes an example of a region R' of the kind we wish to consider. The point C, for instance, is a boundary point as well as an interior point of R'; as a point of the rectangle, it is a boundary point, and as a point of the tongue, it is an interior point. The reader will easily be able to locate points that occur twice as interior points of R'—for example, D. To get a clear picture of things of that kind, it is best to make a paper model of the region. For the time being, the simple example we have just given must suffice in the way of intuitive clarification. In the sequel, we shall call a region *simple* (*schlicht*) if it covers no point more than once; otherwise we shall call it non-simple.

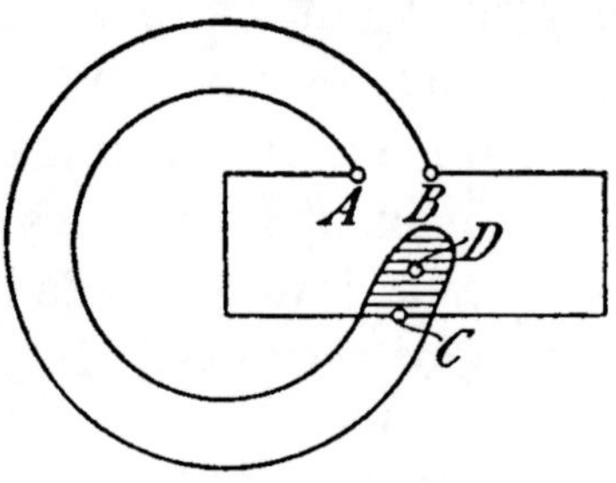

FIG. 1

An *application* of the Theorem on Preservation of Neighborhoods: If $f(z)$ is regular in the interior of a region R, then $|f(z)|$ can not assume a maximum in the interior of the region (*Maximum-Modulus Principle*).

This fact (an easy consequence, as is well known, of Cauchy's Integral Formula) can also be deduced immediately from the Preservation-of-Neighborhoods Theorem. We need only observe that $|f(z)|$ gives the distance of the image point of z from the origin of the w-plane, and that any image point of an interior point of R is the center of a circular disc made up entirely of image points whose pre-images fill up some neighborhood of the above interior point of R in the z-plane; in particular, any point of R' which might claim to be the farthest away from the origin of the w-plane would also be the center of such a disc.

The property of preserving neighborhoods is one that the mappings given by analytic functions share with all mappings that are continuous at every point and one-to-one, or finitely-many-valued, throughout the region. The additional characteristic which singles out the mappings effected by analytic functions, and which is decisive for all our subsequent investigations, is contained in the **Theorem of Isogonality,** or **Preservation of Angles,** which we now proceed to state and prove.

An analytic mapping $w = f(z)$ is *angle-preserving* (or *isogonal*); that is, if $\mathfrak{C}_1$ and $\mathfrak{C}_2$ are any two curves passing through a that are differentiable at a and intersect there at an angle ϑ, then their image curves $\mathfrak{C}_1'$ and $\mathfrak{C}_2'$ in the w-plane intersect, at the image point α of a, at the same angle ϑ (same, that is, as to magnitude and sense of rotation). In order to fully understand the meaning of this theorem, we must keep clearly in mind the hypothesis to the effect that $f(z)$ is regular at the point $z = a$ and has a *non-vanishing derivative* at that point; we must further have a clear agreement as to how the angles that are involved are to be measured. To this end, we first select, once and for all, a positive sense of rotation in the z-plane; this is to be the one that takes the positive half of the x-axis into the positive half of the y-axis in the shortest way. Similarly for the w-plane, u and v taking the place of x and y, respectively. We then assign to each of our two curves $\mathfrak{C}_1$ and $\mathfrak{C}_2$ a definite sense of traversal, and now the angle ϑ between $\mathfrak{C}_1$ and $\mathfrak{C}_2$ at a is defined as that angle through which the (sensed) direction of $\mathfrak{C}_1$ at a must be rotated, in the positive sense, in order to be brought into coincidence with the (sensed) direction of $\mathfrak{C}_2$ at a. Under the given mapping, the sensed curves $\mathfrak{C}_1$ and $\mathfrak{C}_2$ go over into sensed curves $\mathfrak{C}_1'$ and $\mathfrak{C}_2'$. Our theorem now states that *the sensed curve $\mathfrak{C}_1'$ also must be rotated through ϑ, in the positive sense, in order for its direction to be brought into coincidence with that of $\mathfrak{C}_2'$.* This theorem on isogonality is an immediate consequence of the

Cauchy-Riemann differential equations.

Let the two curves be given by $z = z_1(t)$ and $z = z_2(t)$, and let the point $z = a$ correspond to the value $t = 0$ of the parameter, on both curves. Also, let the assigned sense of traversal, for each curve, correspond to increasing t. Assume that the derivatives $z_1'(t)$ and $z_2'(t)$ exist, and that $z_1'(0) \neq 0$ and $z_2'(0) \neq 0$—an assumption which, as is well known, merely serves to exclude singular points of the curves or a poor choice of parameter. Then

$$\vartheta = \arg \frac{z_2'(0)}{z_1'(0)}$$

represents the angle through which the direction of $\mathfrak{C}_1$ at $z = a$ must be rotated, in the positive sense, to be made coincident with that of $\mathfrak{C}_2$ at $z = a$. For if $z' = re^{i\varphi}$, with φ real and $r > 0$, then $\varphi = \arg z'$ is called the *amplitude* (or *argument*) of z'. Hence if $z_1'(0) = r_1 e^{i\varphi_1}$ and $z_2' = r_2 e^{i\varphi_2}$, then

$$\vartheta = \arg \frac{z_2'(0)}{z_1'(0)} = \varphi_2 - \varphi_1$$

is the angle through which $\mathfrak{C}_1$ must be rotated in the positive sense to make its direction coincide with that of $\mathfrak{C}_2$.

The equations of the image curves $\mathfrak{C}_1'$ and $\mathfrak{C}_2'$ are

$$w = w_1(t) = f(z_1(t)), \qquad w = w_2(t) = f(z_2(t)),$$

and for their angle ϑ' we find

$$\vartheta' = \arg\frac{w_2'(0)}{w_1'(0)} = \arg\frac{f'(a)\cdot z_2'(0)}{f'(a)\cdot z_1'(0)} = \arg\frac{z_2'(0)}{z_1'(0)} = \vartheta,$$

which proves isogonality, given that $f'(a) \neq 0$.

We note the following *consequence* of the Isogonality Theorem. Assume a given region R to have a differentiable boundary curve, and let R be mapped on a region R' of the w-plane by means of a function $f(z)$ which is regular in the interior and on the boundary of R. Assign a sense of traversal to the boundary curve $\mathfrak{C}$ of R by fixing the order of three boundary points a, b, and c. The points a, b, c are mapped onto three points a', b', c' on the boundary curve $\mathfrak{C}'$ of R', which we order in the same way as we did a, b, c, thereby assigning to $\mathfrak{C}'$ a definite sense of traversal, which is then said to be induced, or transferred, from that of $\mathfrak{C}$ by means of the mapping. Now if R stays, say, on our left as we traverse $\mathfrak{C}$ in its assigned sense, then similarly R' will stay on our left if we traverse $\mathfrak{C}'$ in the induced sense. This follows directly from the Isogonality Theorem. For if we draw a curve $\mathfrak{K}$ starting at a boundary point a of $\mathfrak{C}$ and pointing into the interior of $\mathfrak{C}$, and denote by α the angle which $\mathfrak{K}$ forms with the assigned direction of $\mathfrak{C}$ at a, then the image curve $\mathfrak{K}'$ of $\mathfrak{K}$ will start at a' and will form the same angle α at that point with the induced direction of $\mathfrak{C}'$. But since

$\mathfrak{R}'$ points into the interior of R', the region R' also lies to the left of its (sensed) boundary $\mathfrak{C}'$.

Next let us verify the fact that our function $f(z)$, regular at $z \neq a$ and with non-vanishing derivative $f'(a)$, gives a **conformal mapping** of a neighborhood of $z = a$. By this is meant that the mapping gives an image which is *similar in the small* to its pre-image, i.e. not only are angles at the point $z = a$ preserved, but so are the ratios of lengths of small segments near $z = a$; to put it precisely, at $z = a$ we also have $dS_1/ds_1 = dS_2/ds_2$, where s_1 and s_2 denote arc-length along $\mathfrak{C}_1$ and $\mathfrak{C}_2$ respectively, and S_1 and S_2, arc-length along $\mathfrak{C}_1'$ and $\mathfrak{C}_2'$ respectively. To prove this, note that $ds_1/dt = |dz_1/dt|$, $ds_2/dt = |dz_2/dt|$, etc., whence it follows that $dS_1/ds_1 = dS_2/ds_2 = |dw/dz|$, Q.E.D.

Note also that dS/ds measures the ratio of magnification of "small" segments at a under the mapping. This "scale factor" of the mapping depends only on the location of a, and not on the direction of the segments; it is the same for s_1 as for s_2.

The theorems on isogonality and isometry which we have just proved have converses in a certain sense; for it can be shown that all isogonal mappings, and likewise all isometric mappings, are given by analytic functions or by functions closely allied to analytic ones.

We shall go into the proof of the first of these statements. Using the notation of p. 1, let us assume that a given mapping $u = u(x, y)$, $v = v(x, y)$ preserves the angle of any pair of curves emanating from $z = a$. Here we assume, as we shall always do in the sequel whenever we speak of mappings, that $u(x, y)$ and $v(x, y)$ have continuous first partial derivatives (are "of class $C^{(1)}$"). We assume here, furthermore, that the functional determinant ("Jacobian")

$$\frac{d(u, v)}{d(x, y)} = \begin{vmatrix} \frac{\partial u}{\partial x} & \frac{\partial u}{\partial y} \\ \frac{\partial v}{\partial x} & \frac{\partial v}{\partial y} \end{vmatrix}$$

is not equal to zero.

[With u and v denoting, as usual, the real and imaginary parts of an analytic function $f(z)$, we have, on account of (1) on p. 1:

$$\frac{d(u, v)}{d(x, y)} = \begin{vmatrix} \frac{\partial u}{\partial x} & -\frac{\partial v}{\partial x} \\ \frac{\partial v}{\partial x} & \frac{\partial u}{\partial x} \end{vmatrix} = \left(\frac{\partial u}{\partial x}\right)^2 + \left(\frac{\partial v}{\partial x}\right)^2 = |f'(z)|^2,$$

so that the non-vanishing of the Jacobian is equivalent with the non-vanishing of $f'(z)$.]

Now a curve $x = x(t)$, $y = y(t)$ is mapped onto $u = u(x(t), y(t))$, $v = v(x(t), y(t))$. The components of the tangent vector to the image curve are then given by

$$u'(t) = \frac{\partial u}{\partial x} x'(t) + \frac{\partial u}{\partial y} y'(t)$$
(3)
$$v'(t) = \frac{\partial v}{\partial x} x'(t) + \frac{\partial v}{\partial y} y'(t).$$

At a fixed point a of the z-plane, the partial derivatives of u and v have fixed values. Thus (3) represents a linear transformation to which the given mapping $w = f(z)$ subjects the tangent vectors to curves in the z-plane at the point a, transforming these tangent vectors into those to the image curves in the w-plane. But as is well known, such a linear transformation preserves angles if and only if it is a similarity transformation. As we know from Analytic Geometry,[1] a necessary and sufficient condition for this is represented, in our case, precisely by the Cauchy-Riemann differential equations (1) on p. 1. But as is shown in Function Theory, these imply that $u + iv$ is an analytic function of z. (Cf. Bieberbach, *Lehrbuch der Funktionentheorie*, Vol. I, p. 39.) We thus have proved the following result:

Every isogonal mapping of a region is represented by an analytic function.

---

[1] In Analytic Geometry, where integral linear transformations of the cartesian coordinates are called *affine* transformations, it is shown that the only angle-preserving affine transformations are the similarity transformations (similitudes).

Let us now go back to our definition of conformal mapping; its requirement of "similitude in the small" includes isogonality, as one part. Let us relax this requirement to the extent that only the magnitude, but not necessarily the sense of rotation, of every angle is to be preserved. A conformal mapping which preserves the sense of rotation will be called *strictly conformal*, and one which reverses the sense of every angle will be called *anti-conformal.* We may then ask whether or not every conformal mapping is given by an analytic function, or—which by our last result amounts to the same—whether or not every conformal mapping is strictly conformal.

A simple example shows that the answer is no. If $\bar{z}$ denotes the complex conjugate of z, then $w = \bar{z}$ represents a mapping of the z-plane onto the w-plane. In geometric terms, this mapping is a reflection in the real axis; that is, if we accommodate z and w both in the same plane in such a way that equal values of z and w correspond to the same point, then the mapping sends every point of the z-plane into its mirror image with respect to the real axis. This mapping is evidently conformal; the length of any curve is the same as that of its image curve, and the *magnitude* of every angle is preserved. However, the *sense of rotation* of every angle is reversed under the mapping. Thus the mapping is anti-conformal. We can obtain additional examples of this kind by combining any given strictly conformal mapping with the mapping just considered.

We may still ask whether every conformal mapping is either a strictly conformal mapping or else a combination of a strictly conformal mapping with the above reflection (and therefore anti-conformal). The answer here is yes, as can be shown by an argument very similar to the one used in proving our last isogonality result, and which we omit here for that reason. Hence *every conformal mapping is represented either by* $w = f(z)$ *(strictly conformal mapping) or by* $\overline{w} = f(z)$ *(anti-conformal mapping), where* $f(z)$ *is an analytic function.* Every conformal mapping preserves the magnitude, and either preserves or reverses the sense of rotation, of every angle.

In what follows we shall be interested only in conformal mappings that are strictly conformal, and that are therefore represented by analytic functions.

§ 2. Integral Linear Functions

The simplest example to illustrate the general discussion of § 1 is furnished by the integral linear functions $w = az + b$. We can distinguish several types among these, as follows:

1. $\boldsymbol{w = z + b}$. If we interpret w and z as

points in the same plane, then this mapping, geometrically interpreted, is a translation. For, as is well known, to every complex number there corresponds a vector, and the addition of complex numbers then corresponds to vector addition. Therefore the translation must be one in the direction of the vector b, and the magnitude of the translation is the length of b. Thus any given region R is mapped onto a congruent region which is obtained from R by a *translation*.

2. $w = e^{i\varphi} \cdot z$ represents a *rotation* of the plane through the angle φ, about the fixed center of rotation $z = 0$.

3. $w = rz$, with $r > 0$, represents a similarity transformation (*magnification* in the ratio $r:1$).

4. The **most general** integral linear transformation $w = az + b$ can be built up step by step from the three types just considered. We set

$$a = re^{i\varphi},\ z_1 = rz,\ z_2 = e^{i\varphi} z_1,\ w = z_2 + b.$$

This shows how the given transformation is built up from the above three types. Another method, just as simple as the one just used, will give us an even better insight into the geometric nature of the mapping $w = az + b$: Observe that it can always be brought into the form $w - \alpha = re^{i\varphi}(z - \alpha)$ with suitable $\alpha, a = re^{i\varphi}$, provided only that the coefficient a in $w = az + b$ is not equal to 1. Therefore *the most general in-*

tegral linear transformation represents either a translation $(a = 1)$, *or a "rotation with magnification"* $(a \neq 1)$, *which reduces to a pure rotation if* $r = 1$ *and to a pure magnification if* $\varphi = 0$.

§ 3. The Function $w = 1/z$

The discussion of this function offers no particular difficulty, at least at all those points at which neither z nor w become infinite, that is to say, at all finite points of the z-plane other than $z = 0$. The point $z = 0$ itself is not covered by the general investigations of § 1; thus if we now include that point in the discussion of our function, we shall at the same time be supplementing the material of § 1 in a special case.

We shall find it useful to introduce polar coordinates, by setting $z = re^{i\varphi}, w = \varrho e^{i\vartheta}$. Then our mapping is expressed by $\varrho = 1/r$, $\vartheta = -\varphi$. This will give us a clear geometric picture of the mapping, as follows. Let us once more locate z and w in one and the same plane. The points for which $r = 1$ obviously play a special role in the mapping. These points make up a circle[1] whose radius is unity and whose center is at $z = 0$, the **unit circle** as we shall henceforth call it for short; and this circle is mapped onto itself under our mapping

[1] By *circle* we shall always mean the periphery of a circular disc.

$w = 1/z$. The point $r = 1$, $\varphi = \varphi_0$ is mapped on the point $\varrho = 1$, $\vartheta = -\varphi_0$, i.e. on the point which is obtained from the first one by "reflecting" the unit circle in the real axis. By the "real axis" we mean the line $y = 0$ $(z = x + iy)$, and by **reflection** in this axis we mean the passing from any given point to its symmetric image,[2] or in other words, the passing from $x + iy$ to $x - iy$. As has already been mentioned in § 1, such a mapping preserves only the magnitude of angles but not[3] their sense of rotation, as can be seen from Fig. 2.

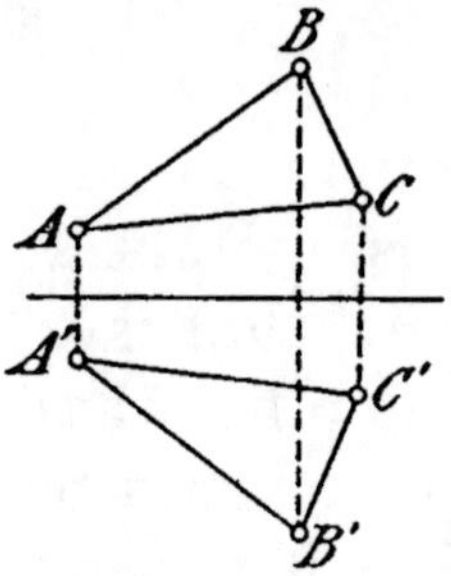

FIG. 2

Let us go on to the consideration of arbitrary z. Clearly, the mapping $w = 1/z$ is one-to-one (excepting the cases $z = 0$ and $w = 0$, which will be discussed later); to every z there corresponds exactly one w, and vice versa. To obtain a clear over-all picture of the mapping, it is useful to decompose it into the following two mappings which are to be applied consecutively:

$$\text{I} \quad r_1 = r, \ \vartheta_1 = -\varphi; \qquad \text{II} \quad \varrho = \frac{1}{r_1}, \ \vartheta = \vartheta_1 .$$

The first of these is simply the reflection in the

[2] It is clear from this what will be meant by reflection in any arbitrary straight line.

[3] We speak of a "reversal" of angles in such cases.

real axis, discussed above, and is therefore an anti-conformal mapping (i.e., conformal mapping which reverses angles). Hence the second transformation, $\varrho = 1/r_1$, $\vartheta = \vartheta_1$, called an ***inversion in the unit circle,*** must also be an anti-conformal mapping, since the combination of the two mappings is a strictly conformal mapping.

Let us investigate the inversion in the unit circle in more detail. We see first that it maps any point $r < 1$, i.e. any point of the interior of the unit circle, onto a point with $\varrho > 1$, i.e. onto a point of the exterior of the circle; and vice versa, it maps the exterior of the unit circle into its interior. The points of the unit circle itself are each left fixed. Thus inversion in a circle interchanges the interior and exterior just as reflection in a straight line interchanges the two sides (half-planes) of the line each of whose points remains fixed. For this reason, the mapping $\varrho = 1/r$, $\vartheta = \varphi$ is also called a **reflection in the unit circle.** Another, and deeper, reason for this terminology will be brought out later in this book. How to find the image under inversion of any given point is made immediately apparent by recalling a familiar theorem on right triangles. Draw the half-line from O through P; the image point P_1 of P must lie on the same line, since $\vartheta = \varphi$. Now if P lies, say, in the interior of the unit circle, then we draw a perpendicular at P on the radius through P, intersecting the circle at T and T', and at these two points we draw the tangents to the circle. P_1 is then the point of intersection of these two tangent lines. If, on the other hand, P_1 is the given point, then its image P is found by simply carrying out the construction we have just described in the reverse direction (see Fig. 3). The validity of the construction rests on the theorem referred to above, according to which $1 = OT^2 = OP \cdot OP_1$.

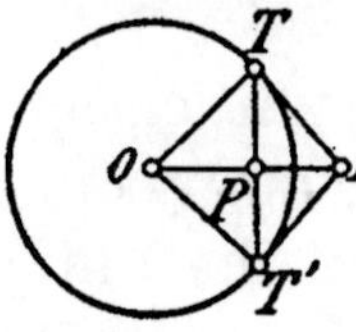

FIG. 3

The above also makes obvious what is to be meant by a **reflection in a circle of radius *R*** about $z = 0$, namely, the mapping $r\varrho = R^2$, $\vartheta = \varphi$.

The geometric meaning of $w = 1/z$ *is now clear: The given point is to be reflected in the unit circle as well as in the real axis in order to arrive at its image point* (**Inversion plus Reflection**).

The mapping is one-to-one, except for the point $z = 0$ to which there does not correspond any image point—at the moment, for we shall presently remove this exception—and except for $w = 0$, which is not—at the moment—to be found among the image points. Now we observe, however, that the exterior of any circle about $z = 0$ as center is mapped into the interior of a circle about $w = 0$, and that the latter shrinks down to $w = 0$ as the radius of the former is made to increase indefinitely. It is just as though there were a point in the z-plane which is *outside every circle* about $z = 0$ and is mapped onto $w = 0$, and as though there were a point in the w-plane which is *outside every circle* about $w = 0$ and onto which $z = 0$ is mapped. The reader may be familiar with a similar state of affairs in Projective Geometry, where one introduces an "improper" or "ideal" straight line, also called the "line at infinity." In our present case, of inversion plus reflection, we introduce a single *improper* point which we denote by $z = \infty$ (or $w = \infty$, respectively). We shall also speak of it as the *point at infinity*. Then $w = \infty$ is the image of $z = 0$ under our mapping, and $z = \infty$ is the pre-image of $w = 0$. With this agreement, we have made $w = 1/z$ a one-to-one mapping, without any exceptions.

Our mapping by reciprocals (i.e., inversion plus reflection) is of great fundamental importance. For, just as one uses collineations in Projective Geometry in order to study the behavior of curves at infinity, so one uses the mapping by reciprocals in Function Theory in order to study the behavior of a function at infinity. We call a function $f(z)$ *regular at (the point at) infinity* if $f(1/w)$ is regular at $w = 0$, so that it can be expanded in powers of w in a neighborhood of $w = 0$. Thus a function regular at $z = \infty$ can be expanded in powers of $1/z$, and such an expansion will be valid in some neighborhood of $z = \infty$, i.e. in a region of the z-plane—such as the exterior of a circle about $z = 0$—which is the pre-image under $w = 1/z$ of a neighborhood of $w = 0$. The mapping $f(z)$ is called isogonal at $z = \infty$ if $f(1/w)$ is isogonal at $w = 0$. The angle formed by two curves at $z = \infty$ is defined to be the angle at which their image curves under $w = 1/z$ intersect in the w-plane at $w = 0$.

The function $w = 1/z$ is also a useful tool when it comes to investigate the points at which a given function $w = f(z)$ becomes infinite. If $f(z)$ does not remain bounded in the neighborhood of, say, $z = a$, then we consider instead of $f(z)$ the function $1/f(z)$; if the latter is one-valued and bounded in a neighborhood of $z = a$, we can write down an expansion of the form

$$1/f(z) = (z - a)^n (a_0 + a_1 (z - a) + \ldots),$$

with $a_0 \neq 0$. Hence we obtain

$$f(z) = \frac{1}{(z - a)^n} \left(\frac{1}{a_0} + b_1(z - a) + \cdots \right).$$

We call $z = a$, in this case, a non-essential singularity, or a *pole*, of $f(z)$. If $n = 1$, then the mapping represented by $f(z)$ is isogonal at $z = a$, in accordance with our agreements above.

Appendix to § 3: **Stereographic Projection**

It is often useful to help logical considerations along by illustrating them, if possible, by means of intuitive or pictorial devices. Thus we will gain by illustrating the introduction of the point at infinity by means of a model which is entirely in the finite domain. The addition of the point at infinity entails the possibility of mapping the plane one-to-one and isogonally onto (the surface of) a sphere. This is done by a mapping called *stereographic projection.* We take a sphere of diameter unity and lay it on the plane in such a way that its lowest point coincides with $z = 0$. This lowest point we call the south pole, and the diametrically opposite point we call the north pole. With the north pole as center of projection, we now project the plane onto the sphere. The points of the sphere and the points of the plane are thereby put in a one-to-one correspondence, under which the south pole, for instance, corresponds to $z = 0$, while the north pole corresponds to the point at infinity of the plane. A short argument will now show that this mapping is isogonal.

By the angle between two plane curves is meant the angle between their tangents at their point of intersection; by the angle between two curves on the sphere is meant the angle between the tangents to the sphere that are also tangent to the curves at their point of intersection. Now let $\mathfrak{C}_1$ and $\mathfrak{C}_2$ be two curves in the plane that intersect at P, and let $\mathfrak{C}_1'$ and $\mathfrak{C}_2'$ be their spherical images (under stereographic projection), intersecting at the image P' of P. Let us pass two planes through the projecting ray PP', containing the tangents at P to $\mathfrak{C}_1$ and to $\mathfrak{C}_2$ respectively; these planes clearly contain also the tangents to the sphere that are tangent at P' to $\mathfrak{C}_1'$ and $\mathfrak{C}_2'$ respectively. The two last-mentioned tangents, in turn, lie in the tangent plane to the sphere at P'. Now pass a meridian plane of the sphere (i.e., a plane containing both the north and the south poles) through PP'—this is the plane in which Fig. 4 is drawn. In Fig. 4, t is the trace of the tangent plane, e is the trace of the z-plane, s is the projecting ray, M is the center of the sphere, and N is the north pole. If we consider the dotted lines drawn in Fig. 4 and recall certain familiar theorems of elementary geometry, we see that t and e form the same angle α with s. Now the two planes which we passed through s are seen to be intersected by two planes, through e and t respectively (viz., the z-plane and the tangent plane at P'), that can be obtained from each other by a reflection in the plane of the perpendicular bisectors of PP'. There-

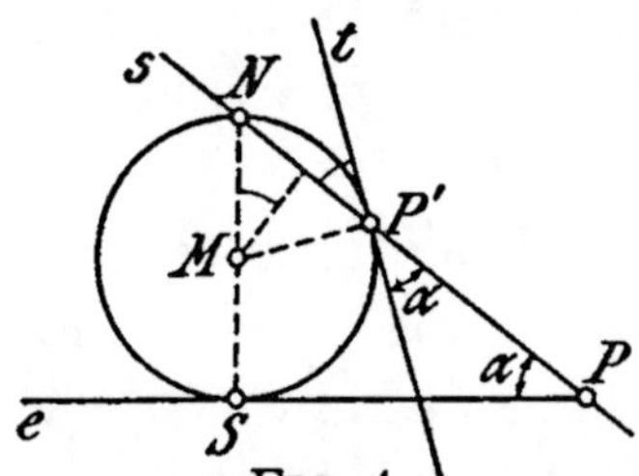

FIG. 4

fore the two pairs of lines of intersection—viz., the two pairs of tangents (at P and at P')—form equal angles, and we have proved that stereographic projection is isogonal at any finite point P; finally, the isogonality of the mapping at the point at infinity, whose image is the north pole, follows from our convention on how to measure angles at infinity.

§ 4. Linear Functions

It would seem natural to begin the investigation of the linear (more properly: fractional linear) function $w = \dfrac{az+b}{cz+d}$ by dividing the denominator into the numerator, which in the case $c \neq 0$ yields

$$w = \frac{a}{c} + \frac{bc - ad}{(cz+d)c}.$$

And then it would be easy to represent our function as built up from four simple types of functions such as were discussed in §§ 2 and 3. But such a procedure would make the further study of the linear function somewhat laborious, and for this reason we shall prefer a different approach. Let us, however, note the following corollary to our initial calculation:

THEOREM I. *The linear function*

$$w = (az + b)/(cz + d)$$

is non-constant if and only if the determinant $ad - bc$ *does not vanish.*

We shall always assume this condition to be satisfied in the sequel. The linear function then has an inverse, which we calculate as $z = \dfrac{dw - b}{-cw + a}$. We deduce from this the following:

THEOREM II. *Every (non-constant) linear function represents a one-to-one mapping of the plane onto itself, and this mapping is isogonal at every point (including* $z = \infty$*).*

That this holds at $z = \infty$ follows from first substituting $z = 1/\mathfrak{z}$ and then noting that $\left(\dfrac{a + b\mathfrak{z}}{c + d\mathfrak{z}}\right) = w$ is regular at $\mathfrak{z} = 0$, except if $c = 0$, and that $\left(\dfrac{dw}{d\mathfrak{z}}\right)_{\mathfrak{z}=0} = \dfrac{bc - ad}{c^2}$ does not vanish. This being so, we are justified in saying that w is isogonal at $z = \infty$, in accordance with our agreement of § 3. But if $c = 0$, we consider $d/(az + b)$ at $z = \infty$, in accordance with § 3. Finally, the isogonality at $z = -d/c$ follows immediately from the fact that $\dfrac{cz + d}{az + b}$ is isogonal at this point.

Let us introduce a few abbreviations. We shall

use S to stand for any linear function, whose inverse—that we have just seen how to calculate above—we shall then denote by S^{-1}, as is usual in algebra. The following further result is now almost immediate:

THEOREM III: *The composition of any number of non-constant linear functions always leads to further non-constant linear functions.*

In proof, let $S_1 = l_1(z)$ and $S_2 = l_2(z)$; then $S_1 S_2$ stands for $l_1(l_2(z))$. The inverse of this last is $S_2^{-1} S_1^{-1}$. The determinant of $S_1 S_2$ is the product of the determinants of S_1 and S_2, and can not vanish since neither of the factors vanishes.

THEOREM IV: *If the z-plane is mapped onto the w-plane by means of a non-constant linear function, then the totality of straight lines and circles of the z-plane is mapped onto the totality of straight lines and circles of the w-plane.*

In proof, note first that the equations of lines and circles can always be written as follows in rectangular coordinates: $\alpha z\bar{z} + \beta z + \bar{\beta}\bar{z} + \gamma = 0$, where α and γ are real, β and $\bar{\beta}$ are complex conjugates, $z = x + iy$, and $\bar{z} = x - iy$. Theorem IV can then be easily verified by simply going through the actual calculations. Under such a mapping, as the calculations would show, a straight line may very well be mapped onto a circle, but never onto any other conic section nor, say, onto a curve of the third order. For example, $w = 1/z$ maps any circle through $z = 0$ onto a circle through $z = \infty$, i.e. onto a straight line; in particular, it maps any straight line through $z = \infty$ and $z = 0$ onto a straight line of the w-plane.

THEOREM V: *Given any three distinct points α, β, γ of the z-plane, and any three distinct points α', β', γ' of the w-plane, there always exists a suitable linear function which maps α, β, γ onto α', β', γ' respectively, i.e. which maps the first triple of points onto the second triple in a given order. Furthermore, the function which accomplishes this mapping is thereby uniquely determined.*

COROLLARY. Since three points determine a circle and since by Theorem IV, circles (including straight lines) are mapped on circles (or lines), Theorem V may also be given the following geometric interpretation: *Any given circle can be mapped conformally onto any other circle in such a way that any three given points of the first circle are mapped onto any three given points of the second.*

Proof of Theorem V. A function such as the theorem requires can obviously be obtained by elimination of $\mathfrak{z}$ from

$$\mathfrak{z} = \frac{z-\alpha}{z-\gamma} \cdot \frac{\beta-\gamma}{\beta-\alpha} \quad \text{and} \quad \mathfrak{z} = \frac{w-\alpha'}{w-\gamma'} \cdot \frac{\beta'-\gamma'}{\beta'-\alpha'}.$$

For, the above maps $z = \alpha, \beta, \gamma$ on $\mathfrak{z} = 0, 1, \infty$, and maps $w = \alpha', \beta', \gamma'$ on $\mathfrak{z} = 0, 1, \infty$. It is also easy to show that the function thus constructed is the only one which satisfies the conditions of the theorem, as follows: If there were two different such functions, say S_1 and S_2, then $S_2^{-1}S_1$ would leave fixed the three points α, β, γ. But then the last mapping would have to leave *all* points fixed, and S_1 and S_2 could not be distinct. To prove this last statement, let us assume that

$$S_2^{-1}S_1 \equiv w = \frac{az + b}{cz + d}$$

leaves fixed the three points α, β, γ; then the quadratic equation

$$z = \frac{az + b}{cz + d}$$

or

$$z^2c + z(d - a) - b = 0$$

must have the *three* solutions α, β, γ. But then all the coefficients of the quadratic equation must vanish, by an elementary theorem of algebra. This gives $b = 0$, $c = 0$, $a = d$. Hence $w = z$ is the only linear function which leaves more than three points fixed. We have thus obtained the following further result:

THEOREM VI. *Every linear function other than $w = z$ leaves at most two points fixed.*

To find the coordinates of these **fixed points,** as we shall call them, we go back to the above quadratic equation, from which we find, in case $c \neq 0$,

$$(1) \qquad z = \frac{a - d \pm \sqrt{(a-d)^2 + 4bc}}{2c}.$$

The two fixed points coincide if

$$(a-d)^2 + 4bc = 0.$$

If $c = 0$, we are dealing with an integral linear function, which leaves $z = \infty$ fixed, and whose finite fixed point is

$$z = b/(d-a).$$

Now to begin a more detailed study, let us first investigate those linear functions that have *two distinct finite fixed points.* Let z_1 and z_2, then, be the two fixed points of $w = \dfrac{az+b}{cz+d} \equiv S$, where z_1 corresponds to, say, the upper sign in (1). In order to be better able to visualize what is going on, we shall again interpret z and w as points in one and the same plane. We shall also use an auxiliary plane in which we accommodate the variables $\mathfrak{w}$ and $\mathfrak{z}$, defined by

$$\mathfrak{w} = \frac{w - z_1}{w - z_2}, \quad \mathfrak{z} = \frac{z - z_1}{z - z_2} \equiv L.$$

The linear function LSL^{-1}, which expresses $\mathfrak{w}$ in terms of $\mathfrak{z}$, has 0 and ∞ as its fixed points, and must therefore be of the form $\mathfrak{w} = a\mathfrak{z}$. This implies that S itself may be written in the form

$$\frac{w - z_1}{w - z_2} = \alpha \frac{z - z_1}{z - z_2}$$ (**Normal Form** in the case of two distinct finite fixed points). In order to determine the value of α in terms of the original coefficients of S, note that $\alpha = \frac{w - z_1}{w - z_2} \cdot \frac{z - z_2}{z - z_1}$, whence a short calculation yields

$$\alpha = \frac{a + d + \sqrt{(a-d)^2 + 4bc}}{a + d - \sqrt{(a-d)^2 + 4bc}}.$$

The relations we have just discussed make it possible for us to restrict ourselves, at least to begin with, to the function $\mathfrak{w} = a\mathfrak{z}$, since we can always pass from this to the general linear function under discussion by making the substitutions

$$\mathfrak{w} = \frac{w - z_1}{w - z_2}, \quad \mathfrak{z} = \frac{z - z_1}{z - z_2}.$$

Three distinct cases now present themselves: 1. If a is a positive real number, our linear function is said to be **hyperbolic**; 2. if $\alpha = e^{i\omega}$ (and a not positive), the function is said to be **elliptic**; 3. all other linear functions with two finite fixed points, and $\alpha = \varrho \cdot e^{i\omega}$, are said to be **loxodromic.**

The geometric meaning of these mappings is easily understood in terms of $\mathfrak{w}$ and $\mathfrak{z}$, recalling that $\mathfrak{w} = a\mathfrak{z}$. The hyperbolic mappings are magnifications, the elliptic ones are rotations, and the loxodromic ones are a combination—referred to on p. 15 above as "rotation plus magnification"—of the first two types. Let us scrutinize the first two types a little more closely. For these, a special role is played on the one hand by the system of straight lines through $\mathfrak{z} = 0$, and on the other hand by the system of circles about $\mathfrak{z} = 0$ as center, as these two families of curves are mapped onto themselves by the two types of mappings. In particular, any *hyperbolic* function maps each of the above straight lines onto itself while permuting the above circles among themselves; whereas any *elliptic* function maps each of the circles onto itself while permuting the straight lines among themselves. It only remains to locate the families of circles which take the place of the above two families when we return to the general case of two arbitrary fixed points instead of the special fixed points $\mathfrak{z} = 0$ and $\mathfrak{w} = 0$ that we have just considered, i.e. when we return from the auxiliary variables $\mathfrak{z}$ and $\mathfrak{w}$ to the original variables z and w. The desired families are clearly those that are obtained from the above two by the mapping $\mathfrak{w} = \frac{w - z_1}{w - z_2}$, $\mathfrak{z} = \frac{z - z_1}{z - z_2}$ in the z, w-plane.

They are the system of circles through the two fixed points and the system of orthogonal trajectories of these circles (see Fig. 5).

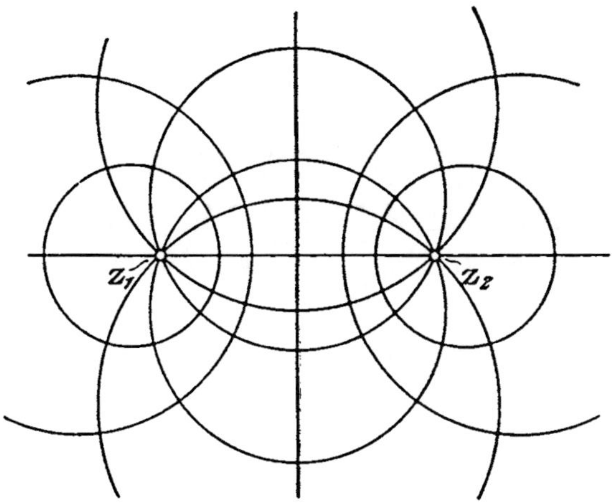

FIG. 5

In the case of loxodromic mappings, systems of circles playing roles as described above do not occur unless $\omega = \pi$.

The circles that remain fixed individually are sometimes called *path curves* or *trajectories*, and those that are permuted among themselves, *level curves* of the linear function. We shall explain the reason for this terminology, which is obviously borrowed from kinematics, in the case of $\mathfrak{w} = a\mathfrak{z}$. If $\mathfrak{w} = a\mathfrak{z}$ is any similarity transformation with $a \neq 1$, then by using a real parameter t we can write all hyperbolic substitutions in the form

$y' = a^t y$. Similarly, we can obtain all rotations from one given rotation. If we interpret t as time, we can see that every point moves along its trajectory as time passes, and that the level curves are changed into each other.

Similarly, we can generate a whole family of mappings out of a single loxodromic mapping; however, not all members of such a family will be loxodromic. If we follow the path of a point as time goes on in this case, we obtain spirals both for the trajectories and for the level curves.

Next, let us study the linear mappings that have only one fixed point z_1; for these, as they constitute a limiting case, we shall use the name **parabolic** mappings. Once more we interpret z and w as points in one and the same plane, and pass to an auxiliary plane by means of the substitution $\mathfrak{w} = \frac{1}{w - z_1}$, $\mathfrak{z} = \frac{1}{z - z_1}$. This yields $\mathfrak{w} = \mathfrak{z} + \beta$, whence we obtain $\frac{1}{w - z_1} = \frac{1}{z - z_1} + \beta$ as the **Normal Form** for parabolic mappings. We also see that the special case $\mathfrak{w} = \mathfrak{z} + \beta$ just considered belongs to the translations, discussed earlier. Under this translation, every straight line that is parallel to the direction $O\beta$ is mapped onto itself, while their orthogonal trajectories form a second system of (parallel) lines, and these are permuted among themselves by the translation. In the

z, w-plane, the role of the two systems of lines just discussed is taken over by two systems of circles passing through z_1. The circles of each system, being conformal images of a system of parallel lines, have a common direction (i.e. a common tangent) at z_1. The fixed point z_1 is the intersection of two mutually perpendicular straight lines, each of which is tangent at z_1 to all the circles of one of the two systems of circles (see Fig. 6).

FIG. 6

§ 5. **Linear Functions** (continued)

We shall investigate next all those linear functions that correspond, under stereographic projection, to **rotations of the sphere.** Any rotation of the sphere maps the (surface of the) sphere isogonally onto itself. Since, furthermore, stereographic projection maps the sphere isogonally onto the plane, it follows that the rotations of the sphere must correspond to one-to-one isogonal mappings of the plane onto itself. In this connection, the following theorem holds:

Every one-to-one isogonal mapping of the plane onto itself is a linear mapping.

For, according to the results in § 1, every such mapping is represented by an analytic function. Let $w = f(z)$ be this function; either it has $z = \infty$ as a fixed point, or it maps $z = \infty$ onto a finite point $w = a$. In the latter case, we form the function $\mathfrak{z} = \dfrac{1}{f(z) - a}$, which represents a one-to-one conformal mapping (from z to $\mathfrak{z}$) that leaves ∞ fixed. Since it assumes each value only once, it can not come arbitrarily close to every value in the neighborhood of $z = \infty$, and must therefore have a pole at infinity. But since it is regular in the entire finite z-plane, it must be an integral rational function, and its degree must be unity, since it assumes no value more than once.

We are now ready to determine all the linear mappings that correspond to rotations of the sphere. Every such mapping must have two fixed points, namely the two points in the plane that correspond to the two intersections of the sphere with the axis of rotation. Where are these two fixed points?

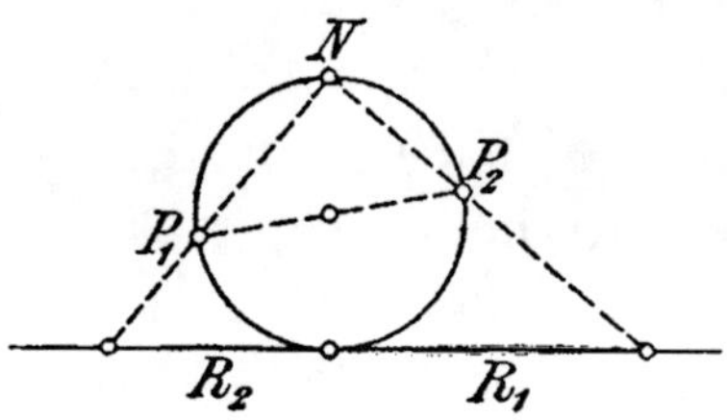

FIG. 7

Let P_1 and P_2 be the two intersections just referred to, and let N be the north pole of the sphere. These three points determine a meridian on the sphere, in whose plane Fig. 7 is drawn. The segments R_1

and R_2 are located on the intersection of this meridian plane with the z-plane. Since P_1P_2 is a diameter (of length unity), the angle at N is a right angle. Therefore $R_1R_2 = 1$, by a familiar theorem on right triangles. From this it follows that if a is one of the fixed points we are seeking, then the other one must be $-1/\bar{a}$. If we observe further that the trajectories of the rotation of the sphere (or more precisely: of the linear mapping that corresponds to that rotation) are the stereographic images of the circles on the sphere that are cut out by planes perpendicular to the axis of rotation, we see that rotations of the sphere give rise to elliptic mappings and that, vice versa, all elliptic mappings with fixed points as described above correspond to rotations of the sphere. Using these facts and solving for w in the normal form for elliptic mappings (cf. the preceding section), we find the following general form for the linear functions corresponding to rotations of the sphere:

$$w = \frac{az + b}{-\bar{b}z + \bar{a}}.$$

We note that this formula could also be derived immediately from the fact that a rotation of the sphere sends any pair of diametrically opposite points of the sphere into another such pair.

As the second example to be worked out, we now choose the **linear mappings of a circular disc onto itself.**

1. As we know from the Corollary to Theorem V on p. 26, any given circle can be mapped onto any other circle by a linear function, and what is more, it is also possible in doing this to prescribe the mapping of any three given points of the given circle. In particular, we can map any given circle onto the real axis (i.e., onto the "circle" through ∞, 0, 1). Under such a mapping, the interior of the circle must be mapped either into the upper half-plane $y > 0$ or into the lower half-plane $y < 0$ $(z = x + iy)$, since conformal mappings preserve neighborhoods. We can always arrange for the upper half-plane, say, to be the image of the interior of the circle, since if necessary we can always use $w = 1/z$ to interchange the two half-planes.

2. Let us now consider, in particular, those linear functions which map the upper half-plane onto itself. To obtain the most general linear function that maps the real axis onto itself, we write down that function which maps the three arbitrary real points α, β, γ onto the three points 0, 1, ∞, using the appropriate formula on p. 26 above.

This yields

$$w = \frac{z-\alpha}{z-\gamma} \cdot \frac{\beta-\gamma}{\beta-\alpha}. \tag{1}$$

This is evidently[1] the most general linear function $w = \frac{az+b}{cz+d}$ with real coefficients. All of these functions map the real axis onto itself, but not all of them map the upper half-plane onto itself; they may interchange the two half-planes, as does, for instance, $w = 1/z$. How can we distinguish between the two cases? We shall show that the upper half-plane is mapped onto itself if and only if $ad - bc > 0$; indeed, this follows from the representation (1) of our functions. For, the determinant of (1) equals $(\beta-\gamma)(\beta-\alpha)(\alpha-\gamma)$. Now for the upper half-plane to be mapped onto itself, it is necessary and sufficient that the order of α, β, γ agree with that of $0, 1, \infty$; but this implies that the value of the determinant is positive, and vice versa. We thus have the following theorem: *All linear conformal mappings of the upper half-plane onto itself are given by* $w = \frac{az+b}{cz+d}$ *with real coefficients and with* $ad - bc > 0$. (This is of course not the most general way of

[1] By p. 26 above, every linear mapping is determined uniquely by the specification of the images of any three given points. Therefore every real linear mapping can be written in the form (1), with real α, β, γ.

writing such mappings; we may, for instance, multiply numerator and denominator by common factors, and these may be non-real complex numbers.)

3. Since we have now learned how to find all linear mappings of a given circular disc onto a half-plane, we can also solve the problem of finding *all linear mappings of any given circular disc onto itself*. We shall only note down the result for the circular disc of radius unity with center at $z=0$, for whose mappings onto itself we obtain

$$w=\frac{az+b}{\bar{b}z+\bar{a}}, \quad a\bar{a}-b\bar{b}>0.$$

The *fixed points* of a linear function that maps the upper half-plane onto itself are either real or complex conjugates, as can be seen from (1) on p. 28. If the fixed points are real, the mapping is either hyperbolic or parabolic or loxodromic with negative multiplier α; if they are complex conjugates, the mapping is elliptic (cf. the calculation of α and of the fixed points on pp. 28 and 29). Hence a loxodromic mapping whose multiplier is non-real can never map a circle onto itself.

In order to obtain similar information concerning the location of the fixed points in the case of functions that map the interior of a *circle* into itself, we need only find out what happens to a pair of points symmetric with respect to the real axis when the upper half-plane is mapped onto a circular disc. The following general theorem contains the answer:

If a linear function maps one circular disc onto another, it maps any pair of points related by inversion in the first circle onto a pair similarly related with respect to the second circle.

This follows from the following remark: If a circle K' is passed through two points P and Q that are mutually inverse with respect to a circle K, then K' and K intersect at right angles. For if K is a straight line, then the center of K' must lie on K, while if K' has radius R, we may first draw the tangents to K' through the center M of K; the square of their length is $= |MP| \cdot |MQ|$, by a well-known theorem of elementary geometry. This last expression, however, has the value R^2, since P and Q are inverses with respect to K. Thus the tangents to K' from M are of length R, and therefore their points of tangency to K' are the points of intersection of K and K', whence K' and K intersect at right angles. Vice versa, the same theorem of elementary geometry that was just used implies that every circle K' perpendicular to K consists entirely of pairs of points mutually inverse with respect to K. Because of the isogonality of linear mappings, they map any circle perpendicular to K onto a circle perpendicular to the image of K, and since any pair of points mutually inverse with respect to K lies on a circle perpendicular to K, the points of the image pair must be mutually inverse with respect to the image of K, which is what we wished to prove.

4. The solution, indicated above under 3., of the problem of determining all *linear* mappings of a circular disc onto itself, acquires an even greater importance through the fact, to be proved presently, that there are *no other one-to-one conformal mappings of a circular disc onto itself.*

To prove this, it will suffice to prove that all one-to-one conformal mappings of the circular disc $|z| < 1$ that leave its center $z = 0$ fixed, are linear. For, any other given point of this disc can be mapped onto $z = 0$ by means of a suitable linear mapping of $|z| < 1$ onto itself, e.g. by means of a suitable hyperbolic function whose two fixed points are the end-points of the diameter on which the given point lies. For the class of functions that leave $z = 0$ fixed, we shall base the proof on the following lemma.

Schwarz' Lemma. *Let* $f(z) = a_1 z + a_2 z^2 + \dots$ *be convergent for* $|z| < 1$, *and let* $|f(z)| \leqq 1$ *for* $|z| < 1$. *Then* $|f(z)| \leqq |z|$ *for all* $|z| < 1$, *with the equality sign not holding for any* $|z| < 1$ *unless* $f(z) = e^{i\alpha} z$ (α *real*).

To prove Schwarz' Lemma, note first that $\frac{f(z)}{z} = a_1 + a_2 z + \cdots$ likewise converges for $|z| < 1$.

Hence in accordance with the maximum-modulus principle mentioned on p. 5, the function $\frac{f(z)}{z}$ can not have a maximum of its modulus occurring in the interior of the circular disc $|z| \leqq \varrho < 1$, whence $\left|\frac{f(z)}{z}\right| \leqq \frac{1}{\varrho}$ for $|z| \leqq \varrho$. This holds for every fixed z and any $\varrho < 1$ that satisfies $|z| \leqq \varrho$. But this implies that $\left|\frac{f(z)}{z}\right| \leqq 1$ for every $|z| < 1$, i.e. that $|f(z)| \leqq |z|$. Now if the equality sign holds for any point $z = a$ with $|a| < 1$, so that $|f(a)| = |a|$, then unless $\frac{f(z)}{z}$ is a constant, the function $\frac{f(z)}{z}$ must map a neighborhood of $z = a$ on a neighborhood of $\frac{f(a)}{a}$, by the preservation-of-neighborhoods theorem. But since $\left|\frac{f(a)}{a}\right| = 1$, this would imply that there are points arbitrarily close to $z = a$ at which $\left|\frac{f(z)}{z}\right| > 1$. This would contradict what we have proved in the first part of this argument. Hence if the equality sign holds anywhere in $|z| < 1$, then $\frac{f(z)}{z}$ must be a constant. In that case, $\left|\frac{f(z)}{z}\right| = 1$

everywhere, and therefore $f(z) = e^{i\alpha}z$ everywhere, which completes the proof of Schwarz' Lemma.

From this we now derive the following in short order:

All one-to-one conformal mappings of the interior of the unit circle onto itself are linear.

For if $w = f(z)$ is such a mapping that in addition leaves $z = 0$ fixed, then according to Schwarz' Lemma, the mapping sends every point into an image point that is at least as close to the origin $z = 0$ as is the original point, and the same holds of course for the inverse mapping. These two facts are compatible only if the mapping does not change the distance from $z = 0$ of any point in $|z| < 1$. But then, again by Schwarz' Lemma, we must have $w = f(z) = e^{i\alpha}z$, which is a linear function, and the proof is through in case $z = 0$ was fixed under the mapping. In case $f'z)$ does not leave $z = 0$ fixed, then a suitable linear function of $f(z)$ will do so (cf. the remark preceding Schwarz' Lemma), and our proof is finished.

Remark. The hypothesis of "one-to-one"-ness is essential to the validity of the theorem just proved, as the example $w = z^2$ (which maps the unit circle onto itself) shows.

Exercises. 1. Given two circular annuli, the first formed by two eccentric circles and the second by two concentric ones; find a hyperbolic linear function that maps the first annulus onto the second.

2. Find the most general "triangle" formed by circular arcs that is mapped onto an ordinary (straight-line) triangle by $w = 1/(z - a)$.

3. Find a function that maps a crescent, formed by two mutually tangent circles, onto an infinite strip bounded by two parallel straight lines.

§ 6. Groups of Linear Functions

By a *group of linear functions* is meant a set of linear functions such that the composition $S_1 S_2$ of any two functions S_1 and S_2 of the set is itself an element S_3 of the set, and such that the set also contains the inverse function S^{-1} of any function S that belongs to the set. (Cf. the notation introduced on p. 25.)

We shall determine a **fundamental region** of such a group. By a fundamental region is meant a region of the following sort: If all the mappings contained in the group are applied, one after the other, to such a region, then the totality of image regions thus obtained should constitute a simple covering either of the whole plane or of a part thereof, and the region should not be a proper subregion of a larger one that also has the covering property just described.

A few examples will serve to illustrate these

definitions. Consider, for instance, a group of rotations about the point $z = 0$. Let the mappings contained in this group be the following ones: $z' = e^{\frac{2hi\pi}{n}} z$ $(h = 1, 2, \ldots n)$, where n is an integer. Thus the group consists of the rotations about $z = 0$ through the angle $2\pi/n$ and the repetitions of this rotation. We see immediately that the composition of two rotations $z' = e^{\frac{2h_1 i\pi}{n}} z$ and $z'' = e^{\frac{2h_2 i\pi}{n}} z'$ of the group yields the rotation $z'' = e^{\frac{2(h_1+h_2)i\pi}{n}} z$ of the group. Furthermore, the rotation inverse to $z' = e^{\frac{2hi\pi}{n}} z$ is $z = e^{\frac{2(n-h)i\pi}{n}} z'$, which is itself contained in the group. As a fundamental region of this group we may take, say, the sector bounded by two rays emanating from $z = 0$ that form the angle $2\pi/n$ at $z = 0$, one of the two bounding rays being included in the region. For if all rotations of our group are applied to this region, we obtain a complete covering of the whole plane by n sectors. Or in other words: Every point of the plane can be mapped, by a suitable rotation of the group, onto a point of the fundamental region, so that for every given point of the plane the fundamental region contains exactly one corresponding point (provided only that one of the bounding half-lines is counted as belonging to the region while the other one is not, as was specified above). If we restricted ourselves to that part of the sector that lies inside the unit circle, we would no longer have a fundamental region, even though the rotations of the group applied to this part would lead to a covering of a portion of the plane (viz., of the unit circle) by congruent circular sectors; and the reason is simply that a region properly containing the finite sector (viz., the whole infinite sector) also gives rise to a simple covering. Nor can we use a sector with double the above angle at $z = 0$ as a fundamental region, since the rotations of the group applied to such a region would yield a covering of the plane which, to be sure, is complete, but which is a double instead of a simple covering. This much must suffice here in the way of an explanation of our definition.

We note further that *the fundamental region of a group is by no means uniquely determined by the group.* We can find quite diverse fundamental regions belonging to one and the same group. Above, for instance, we can replace the angular sector bounded by two straight half-lines with a sector bounded by any two curves that lead from zero to infinity without self-intersections and such that one of them is obtained from the other by the rotation $z' = e^{\frac{2i\pi}{n}} z$.

We also note that not every group of linear mappings need have a fundamental region; for example, the group of *all* rotations about the point $z = 0$ does not have a fundamental region, nor does the group of all those linear functions that have a non-vanishing determinant, as we shall deduce from the following remark: A fundamental region, by its very definition, can not contain two points one of which is obtainable from the other by a mapping belonging to the group, for this would contradict the requirement of obtaining a *simple* covering. Now in the above examples, any given point can be moved to a different one as close to the given one as we please, by a suitable mapping in the group. Thus the fundamental region could not contain any interior points, since an interior point would have to be the center of some circular disc that contains no images of P under any mappings in the group.

The above considerations contain a necessary condition for a group to have a fundamental region, namely that there should be regions containing no pair of points one of which is the image of the other under some mapping in the group. This condition also turns out to be sufficient, as can be seen by enlarging as much as possible some initial region which is free of pairs of points of the kind just described. We shall not carry through the details of such a construction, as this would lead us too far afield here. We shall, however, give a few more examples of groups and their fundamental regions:

1. The group of mappings $w = z + h$, where h is an integer, has as a fundamental region a strip of width unity, bounded, say, by two parallels to the imaginary axis.

2. $z' = z + 2h_1 + 2h_2\omega$ (where h_1 and h_2 are integers and ω is a non-real complex number). As a fundamental region we may take a parallelogram two of whose sides are the vectors joining the origin to 1 and ω.

3. $z' = \frac{1}{z}$, $z' = e^{\frac{2i\pi}{n}}z$ and their composite mappings, with n an integer. This is a so-called *dihedral group*. A fundamental region is the circular sector with its vertices at the origin, at $e^{\frac{i\pi}{n}}$ and at $e^{-\frac{i\pi}{n}}$; the boundary is made up of the arc of the unit circle through $z = 1$ that connects the last two vertices, and of the two radii from $z = 0$ to $e^{\frac{i\pi}{n}}$ and to $e^{-\frac{i\pi}{n}}$

4. The groups of rotations of the other regular solids. Note that under stereographic projection (cf. pp. 21-22), the groups in 3. above correspond to groups of rotations of the sphere that map dihedra onto themselves, these dihedra being double pyramids whose "points" are at the north

and south poles of the sphere. The octahedron, one of the five "regular solids," is among these dihedra. Now the remaining regular solids similarly give rise to groups of rotations, each such group consisting of all rotations that bring the corresponding regular solid into self-coincidence. To find the fundamental regions of these groups, it is best to locate first the corresponding regions on the sphere and then pass to the plane by stereographic projection. To locate the regions on the sphere, however, one proceeds as follows: The triangular faces of the given regular solid are projected onto the surface of the sphere, with the center of the sphere as the center of projection; in each spherical triangle thus obtained, the altitudes are drawn from each vertex to the common point of intersection of the altitudes. The new spherical triangles thus constructed are then stereographic images of fundamental regions of the group associated with the given regular solid. (The cube and the dodecahedron may be omitted, since their groups are identical with those of the octahedron and icosahedron, respectively.)

5. The covering obtained from a fundamental region by the application of the mappings in the group need not be a covering of the whole plane, as it was in the above examples. It may be a covering of some part of the plane only, such as the interior of a circle, or the upper half-plane.

The latter, for instance, is mapped onto itself by the elliptic modular group, consisting of the substitutions $z' = \dfrac{az + b}{cz + a}$, where a, b, c are rational integers satisfying $ad - bc = 1$. A fundamental region for this group, shaded in Fig. 8, is the part outside the unit circle of the strip between the two lines $x = -1/2$ and $x = +1/2$ $(z = x + iy)$.

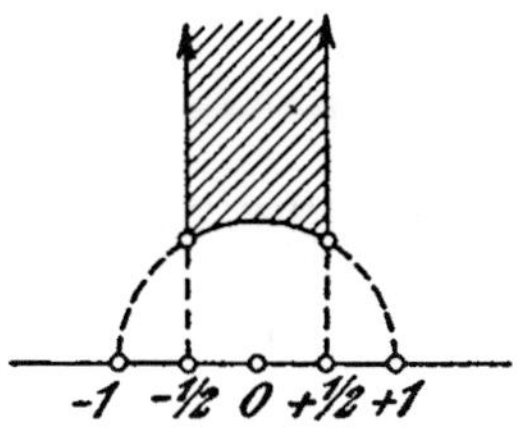

FIG. 8

For the proof, we refer the reader to more detailed expositions (such as Vol. II of the author's *Lehrbuch der Funktionentheorie*, Chelsea Publ. Co., New York 1945). Here we merely add that all the mappings of the group can be generated by composition from two of them, namely from the parabolic mapping $z' = z + 1$ and the elliptic mapping $w = -1/z$. The former maps one of the two boundary lines of the strip onto the other, while the latter has i and $-i$ as its fixed points and maps the two arcs of the unit circle from $+i$ to $-\frac{1}{2} + \frac{i}{2}\sqrt{3}$ and from $+i$ to $+\frac{1}{2} + \frac{i}{2}\sqrt{3}$ onto each other.

An important branch of modern Function Theory is the theory of **automorphic functions.**

It is concerned with functions that remain unchanged under groups of linear functions, in other words, with functions $f(z)$ that satisfy all functional equations $f(z) = f(l_i(z))$, where the $l_i(z)$ represent all the mappings of the given group of linear functions. In simple cases it is easy to find such functions. For example, $w = z^n$ remains unchanged by the rotations $z' = e^{\frac{2hi\pi}{n}} z$. Similarly,

$w = z^n + \frac{1}{z^n}$ is an automorphic function of the

dihedral group of example 3. above. Automorphic functions of the group of Example 2. are given by the elliptic functions; of the group of Example 1., by the function $w = e^{2i\pi z}$; of the group of Example 5., by the elliptic modular function; of the group of Example 4., by functions of the form

$w = \sum_{1}^{n} r(l_i(z))$, where $r(z)$ is a suitable function, the $l_i(z)$ are the mappings in the group, and n is the number of mappings, i.e. the order of the group.

CHAPTER TWO

Rational Functions

§ 7. $w = z^n$

In § 1 we found it necessary to exclude from our discussion, temporarily at least, the singularities of the functions we studied, as well as those of their inverses. In § 3, where we studied the function $w = 1/z$, we took the first step toward closing that gap, and we were able to extend our results in that connection to any function having simple poles only. We shall now take up the function $w = z^n$, whose study will require us to master a new situation.

At $z = 0$, the derivative of the function $w = z^n$ vanishes. The inverse function is not regular at this point; its singularity at $z = 0$ is a so-called **branch-point** of order n. To get a picture of how the mapping $w = z^n$ behaves at $z = 0$, we introduce polar coordinates by setting $z = re^{i\varphi}$, $w = \varrho e^{i\vartheta}$. Then $\varrho = r^n$, $\vartheta = n\varphi$. Thus every circle $r =$ const. is mapped under $w = z^n$ onto a circle $\varrho =$ const., and every straight line $\varphi =$ const. is mapped onto a straight line $\vartheta =$ const. Now it will be convenient to do what the relation $\vartheta = n\varphi$ suggests doing, namely to consider at first only a part of the w-plane, viz., the sector $r > 0$, $0 < \varphi < 2\pi/n$. Its vertices are $z = 0$ and $z = \infty$, and it is bounded by the lines $\varphi = 0$ and $\varphi = 2\pi/n$. This sector, it now turns out, is mapped onto the full w-plane; for, with r and φ ranging over the sector, ϱ and ϑ can independently take on any values whatsoever. The lines $\varphi = 0$ and $\varphi = 2\pi/n$ are both mapped onto the real axis of the w-plane. This discussion gives us an insight into the special nature of the point $z = 0$; the mapping is not isogonal at this point; rather it changes every angle at $z = 0$ into its n-fold in the w-plane, at $w = 0$. For if two curves of the z-plane pass through $z = 0$ with their tangents there intersecting at an angle α, then the tangents to their image curves at $w = 0$ intersect at the angle $n\alpha$. The same holds, as we can see by referring to our above sector, at $z = \infty$; at this point, too, every angle is mapped onto its n-fold. In particular, it follows that the image of our sector covers the whole w-plane. But if the image of only the n-th part of the z-plane covers the whole w-plane, where can the image of all the rest of the z-plane be accommodated? We have no choice but to cover the w-plane a second time, then a third, etc., as often (viz., n times) as necessary. And indeed, the neighboring sector in the z-plane, bounded by $\varphi = 2\pi/n$ and $\varphi = 2 \cdot 2\pi/n$, is also mapped onto a whole w-plane by $w = z^n$. In this way we obtain, corresponding to the n sectors in the z-plane, n full coverings of the w-plane; every point of the w-plane thus appears as the image of n distinct points of the z-plane, and these n pre-images are furnished by the n values of the inverse function $z=\sqrt[n]{w}$. As the above discussion shows, these n values all lie on a circle about $z=0$ as center in the z-plane, and they constitute the vertices of a regular n-gon. Only the points $w=0$ and $w=\infty$ are exceptional, in that each of these has only a single point of the z-plane as its pre-image, viz., $z=0$ and $z=\infty$, respectively. These two points, then, may be said to be part of all n coverings of the w-plane. We shall interpret each separate covering of the w-plane as filling out a separate "sheet" of the w-plane, a device that goes back to Riemann. We think of the n sheets corresponding to the n sectors as lying one on top of the other, so that the n points of the n sheets associated with any given value of w lie vertically above each other. The sheets being arranged in the same order (vertically) as are the corresponding sectors (cyclically), we shall fasten each sheet to the next in a manner we shall now describe in detail. To facilitate this description, we shall distinguish two "banks" of the positive real axis of the w-plane, viz., a right one and a left one. The right bank is the image, under the mapping of the sector bounded by $\varphi = 0$ and $\varphi = 2\pi/n$, of the line $\varphi = 2\pi/n$, and the left bank is the image of the line $\varphi = 0$. When the next sector is subjected to the mapping (the one that borders on the first sector along $\varphi = 2\pi/n$, i.e. No. 2 in Fig. 9), the line $\varphi = 2\pi/n$ is mapped onto the left bank and the line $\varphi = 2 \cdot 2\pi/n$ onto the right bank. Any given point of the line $\varphi = 2\pi/n$ goes into two opposite points of the two banks under the mapping of the two sectors bordering on this line, equal values of w being associated with the two image points. Now we shall think of the two banks of the positive real axis that correspond to the common boundary of the two sectors as being joined together, point by point, in the same way as the two sectors hang together along $\varphi = 2\pi/n$. We thus obtain a region which gives a double covering of the w-plane. We proceed with the remaining sectors in the same way in which we just treated the mapping of the second sector; we think of the corresponding sheets overlying the w-plane as being joined together along edges corresponding to boundaries common to adjacent sectors. If there are only two sectors altogether, as is the case with $w = z^2$, we must think of the two banks that still remain free in Fig. 10 as being joined together. The fact that this can not be done without introducing self-intersections of the resulting surface may be a practical difficulty in the construction of a model, but it should not be a stumbling-block to our intuitive visualization of that construction. As is usual also in other contexts in the Theory of Surfaces, we shall here count the curve of self-intersection as two different curves of the surface, having nothing to do with each other except for their incidental coincidence in a drawing, or on a paper model, whose construction the reader is urged to undertake. In the general case, we thus obtain a surface of n sheets, a so-called **Riemann surface.** Two points, called

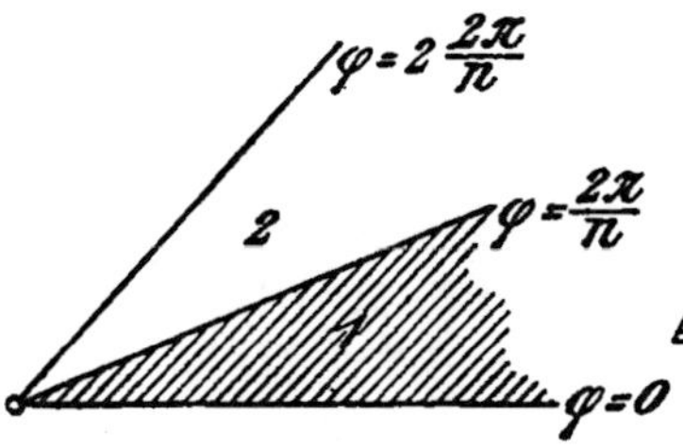

FIG. 9

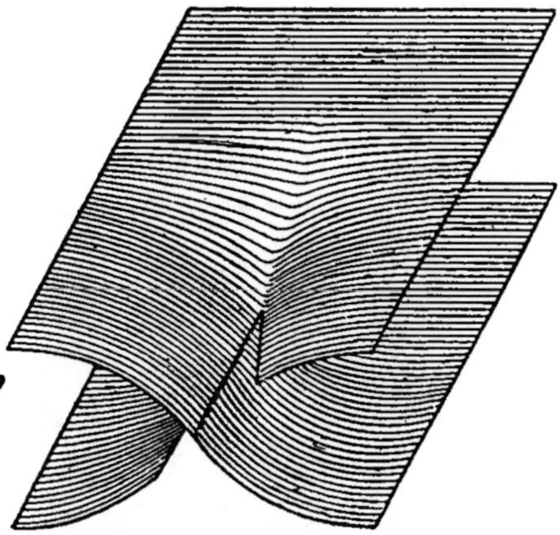

FIG. 10

the **branch points** of our Riemann surface — viz., $w = 0$ and $w = \infty$ — are common to all n sheets. The fact that the sheets were joined together along the *real axis* is only incidental; all kinds of different systems of cuts could be used to give a decomposition of the Riemann surface into n sheets each of which covers the w-plane. Such different systems would correspond to decompositions of the z-plane into sectors different from the ones used above.

A few pages back we saw how the content of the isogonality theorem must be modified to fit the function under discussion (viz., $w = z^n$) at $w = 0$ and $w = \infty$. Let us now ask, what about the Preservation-of-Neighborhoods Theorem in connection with our function? It obviously applies without any modification to the neighborhood of any point other than zero and infinity. Peculiarities are encountered, however, when we deal with the mapping of neighborhoods of $z = 0$ and $z = \infty$, or with the mapping of the whole z-plane. To be sure, the Riemann surface is a closed, connected point set; but it is not a region in the sense in which this term was defined in § 1. It is not possible to describe a circle about $w = 0$ as center and whose interior, covered *simply*, is a sub-region of the surface. However, if we take an n-tuply covered circle with its center and branch-point at $w = 0$, then we are dealing with a sub-region of the surface. This leads us to an **extension of the concept of region:** Henceforth we shall include among the interior points of a region, points near which the region behaves as does our surface near $w = 0$. In other words, any given interior point will have, among its neighborhoods, a simply or multiply covered circle whose center, and only branch-point (if any), is the given point. Thus it will always be possible to map a neighborhood (i.e. all points within some small enough distance) of any interior point a of a region on the surface onto a simple and full neighborhood of $z = 0$ by means of a function $z = \sqrt[n]{w - a}$, where n is a suitable integer, and we adopt this as a necessary and sufficient condition for calling the given point a an interior point of the surface. In the neighborhood of $w = \infty$, the new definition just given must of course be modified, in accordance with our conventions in § 3, in so far as the function $z = \sqrt[n]{w - a}$ is to be replaced by the function $z = \sqrt[n]{\frac{1}{w}}$.

All this being established, we can now say that the Preservation-of-Neighborhoods Theorem remains fully valid for the function $w = z^n$.

The specific Riemann surface considered above is called the Riemann surface of the function $z = \sqrt[n]{w}$ because it is suitable for giving us a geometric picture in the large, so to speak, of the mapping represented by $z = \sqrt[n]{w}$. For if we label every point of the Riemann surface with the z-value of its pre-image under the conformal mapping $w = z^n$, then we have defined a single-valued function $\sqrt[n]{w}$ on the Riemann surface, whereas that same labelling applied to the w-plane as such (instead of to the Riemann surface) defines an n-valued (instead of a single-valued) function. If we observe the values assumed by z when we continue along a curve on the surface by means of the well-known process of analytic continuation, we see that they correspond to our above labelling of points on the Riemann surface. If the function is continued around a closed curve on the Riemann surface, its value will return, upon one full traversal of the closed curve, to its initial value—a statement which will not always be valid if we replace the surface by the simply-covered w-plane. To each of the n values that $z = \sqrt[n]{w}$ assumes for one and the same value of w, there corresponds exactly one point on one of the n sheets of the Riemann surface.

We shall see just below how to construct Riemann surfaces for **more general functions,** to serve the purpose of furnishing a compact picture of how the various branches of a given function are connected. Right now, let us give yet another description of how the construction of the

Riemann surface in the above example can be carried out. We take n duplicates (sheets) of the w-plane and slit each of them open by a cut along the positive real axis from zero to infinity. On each one of the sheets we then accommodate the values of one of the branches of our root function, different sheets being used for different branches; this we can do, say, as follows: We first label n interior points on the n sheets, say $w = i$ on each of them, with the n values that the function assumes there, so that each value is associated with a separate sheet. We then imagine the individual branches of the function to be continued analytically, from the interior points chosen, over the individual sheets, as far as this is possible without crossing the cut on each sheet. In this way we distribute the n branches of the function over the n sheets. We then think of the sheets as being joined along their cuts whenever the function assumes equal values along two edges representing the positive real axis. This process of joining completes our construction of the Riemann surface of the function $z = \sqrt[n]{w}$.

The detailed study of the function $w = z^n$ which we have just made, enables us in the case of other functions also, to see exactly what happens to the conformal mapping in the neighborhood of a zero of the derivative. Let, for example, $w - b =$

$a_n(z-a)^n + \dots$, where $a_n \neq 0$, be such a function. To see how the mapping behaves at $z=a$, $w=b$, we introduce an auxiliary variable t by setting $w-b=t^n$. We find $t^n = a_n(z-a)^n + \dots$, whence

$$t=\beta_1(z-a)+\beta_2(z-a)^2+\cdots.$$

Here, $\beta_1 = \sqrt[n]{a_n} \neq 0$. Solving for $z-a$, we find

$$z-a=\alpha_1 t+\alpha_2 t^2+\cdots.$$

Thus both z and w are one-valued functions of t in the neighborhood of $t=0$, where both are regular. The mapping $w-b=a_n(z-a)^n+\dots$ has thereby been carried out in two steps, namely a first mapping $t=\beta_1(z-a)+\dots$ that maps a simple neighborhood of $z=a$ onto a simple neighborhood of $t=0$, and a second mapping $w-b=t^n$ that maps our simple neighborhood of $t=0$ onto a surface of n sheets winding about $w=b$ as a branch point. Altogether, a simple region of the z-plane is being mapped onto an n-sheeted region, with a branch point, of (or "over") the w-plane. Taking into account our extension on p. 56 of the definition of a region, we may now state that *the Preservation-of-Neighborhoods Theorem applies to all functions that are regular to within poles.*

The facts just discussed may also be interpreted in a slightly different way. Because of the role it has just been shown to play, the auxiliary variable t is called a **local uniformizing variable** for the functional relation represented by $w - b = a_n(z-a)^n + \dots$, since both z and w are one-valued functions of t in the neighborhood of $z = a$, $w = b$. This uniformizing variable plays a role in function-theoretic problems similar to that played by time in problems in mechanics, where the quantities involved in even the most complicated processes of motion may be regarded as single-valued functions of time; the only difference is that in the mechanics problem, the coordinates are single-valued functions of time *at all times,* whereas in our function-theoretic problem, z and w are in general single-valued functions of t *in a sufficiently small neighborhood of* $t = 0$ only. Later on we shall touch on the important problem of representing a functional relation $f(z, w) = 0$ in its entirety by means of parametric equations $z = z(t)$, $w = w(t)$, where $z(t)$ and $w(t)$ are single-valued functions of t. What was done above by means of the local uniformizing variable t, was to map onto a simple neighborhood of $t = 0$ a neighborhood of the place $w = b$ on the Riemann surface of the function $z(w)$ defined by $w - b = a_n(z-a)^n + \dots$. By this means we were able to express w and z as single-valued functions of t **in the neighborhood of $t = 0$.** The solution of the general problem of complete *uniformization in the large* of a given function will depend on the possibility of mapping the **entire** Riemann surface onto a simple region of a t-plane. If this can be done, then all functions that are single-valued on the Riemann surface—in particular, z and w—can be expressed everywhere as single-valued functions of t in the above region of the t-plane.

§ 8. Rational Functions

To supplement our investigations, we shall here study some rational functions from the point of view of how the mappings represented by them behave "in the large." Consider first an integral rational function $w = f(z)$ of degree m; any given value will be assumed by w at m points of the z-plane. Thus in order to obtain the complete image region of the z-plane, we need m sheets of the w-plane, and we must then join these m sheets in the proper way to obtain a Riemann surface. The best way to approach this problem, from the point of view of developing a systematic theory, is to first determine all those values of w that could be associated with branch points of the Riemann surface. These are the points at which the mapping fails to be isogonal, and to find them we must set $f'(z) = 0$. We must further check, by the familiar methods, on the multiple poles (i.e. those of an order greater than unity) and on $z = \infty$. In the w-plane we mark the corresponding values of w. At each of these, the inverse function $z = \varphi(w)$ may have a branch-point of some sort or other. Now we number the actual branch-points in some definite way and then join them in order, starting with the first one, by a continuous, differentiable curve $\mathfrak{C}'$ that does not intersect itself. We then think of the w-plane as being cut along this curve. (In the example $w = z^n$, we had cut the w-plane from 0 to ∞ along the real axis.) In the w-plane as cut in this manner, the function $\varphi(w)$ is single-valued, and it is regular, except for poles, at every interior point of the region bounded by our curve. (In general, $z = \varphi(w)$ will of course assume different values at opposite points of the two banks of our cut.) In order to find out how the n sheets of the Riemann surface must be connected, we first determine the pre-image $\mathfrak{C}$ of $\mathfrak{C}'$ in the z-plane. $\mathfrak{C}$ will consist of several curves. (In the example $w = z^n$, they were the straight lines $\varphi = 2hi\pi/n$ from 0 to ∞.) $\mathfrak{C}$ divides the z-plane into n regions, since to every value of w there correspond n values of z. Whenever two of these regions in the z-plane are contiguous along a part of $\mathfrak{C}$, the two corresponding sheets of the Riemann surface must be connected along the corresponding part of $\mathfrak{C}'$. The surface obtained in this way is closed and is the Riemann surface of the function $z = \varphi(w)$, which is single-valued on this surface and maps it onto the z-plane one-to-one, and, except at the branch-points, isogonally.

One fact that should be stressed here is that not every point that was marked above is of necessity a branch-point on every sheet.

The procedure just outlined may often lead to tedious lengths if all the details are carried through, but it is usually quite adequate to give a schematic idea, so to speak, of what the mapping looks like. Let us take up an *example.* Suppose we are given that

$$w = z^3 + 3z^2 + 6z + 1.$$

We see that $w = \infty$ is a branch-point of order 3. To locate the points where $w' = 0$, we must solve the equation

$$3z^2 + 6z + 6 = 0.$$

This yields $z = -1 \pm i$. We find the following branch-points in the w-plane: $\alpha = -3 - 2i$, $\beta = -3 + 2i$.

As the curve $\mathfrak{C}'$ connecting the three branch-points, we choose the straight line leading from α to ∞ and from there to β. We must then determine the corresponding curve $\mathfrak{C}$ in the z-plane; it consists of the straight line through $-1 + i$, $-1 - i$, and ∞, and the hyperbola having this line as its transverse axis and having its vertices at $-1+i$ and $-1-i$. This can be verified by separating real and imaginary parts.

As a parametric representation for the straight line $\mathfrak{C}'$ in the w-plane, we take $w=-3+it$, where the real parameter t must be $\geqq 2$ in absolute value. For the corresponding curve of the z-plane, we find

$$-3=x^3-3xy^2+3x^2-3y^2+6x+1,$$

$t=3x^2y-y^3+6xy+6y$. We see that the straight line $x=-1$ is part of this curve. The curve being of third order, there remains a conic section which is seen to be the hyperbola described above. Its equation is $3y^2-(x+1)^2=3$.

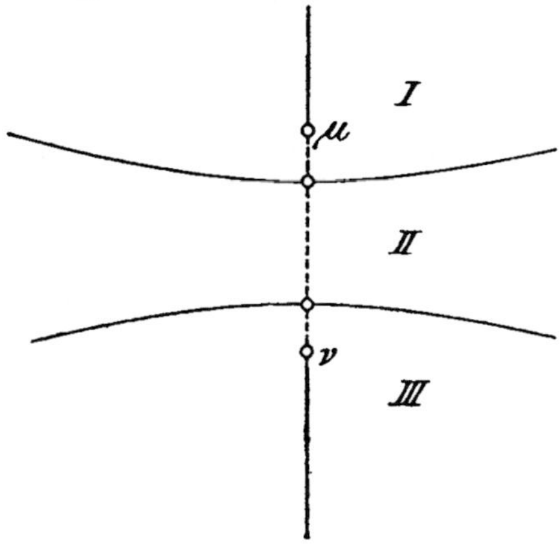

FIG. 11

In Fig. 11, we have numbered with Roman numerals the regions that correspond to the three sheets of the Riemann surface.

The dotted segment of the line $x = -1$, between $\mu = -1 + 2i$ and $\nu = -1 - 2i$, corresponds to the finite segment between α and β of our straight line in the w-plane, as can easily be seen from $t = y(3 - y^2)$, and therefore does not correspond to any part of the system of cuts. Only the two branches of the hyperbola play an essential role in the construction of the Riemann surface, since the two half-lines each lie entirely within one of the three regions and therefore are not part of the common boundary of two different sheets; in joining the Riemann surface together, one merely closes up a cut within a sheet if that cut corresponds to one of the two half-lines just mentioned. Sheets I and II of the Riemann surface are joined along the half-line from $-3 + 2i$ to ∞ of the w-plane; sheets II and III, along the half-line from $-3 - 2i$ to ∞. All the remaining cuts are to be closed up within the individual sheets where they occur. This completes the construction of the Riemann surface.

We shall see presently that the example we have just treated is typical of the mappings represented by integral rational functions of degree three. For, all Riemann surfaces associated with such functions must have a third-order branch-point at infinity, and the finite branch-points are determined from a quadratic equation in z. Hence the three values of w which are possible branch-points always lie on a straight line in the w-plane. Under the inverse mapping, back into the z-plane, this straight line (drawn on the three sheets of the Riemann surface) goes into a curve of the third order, as can easily be seen by separating real and imaginary parts. The straight line, however, passes twice through each finite branch-point, and any such branch-point must go into a double point (point of self-intersection) of the third-order curve. This curve must therefore have two finite double points. The line joining these two double points thus intersects the curve in at least four points and must therefore itself belong entirely to the curve. The curve then is made up of the connecting line plus a conic section which turns out to be a hyperbola except in the case where the two finite branch-points happen to coincide, in which case the hyperbola degenerates into two straight lines. This case is actually realized for the integral rational function $w = z^3$; in all other cases, one obtains a Riemann surface of the same structure as the one we have discussed in this section, except that of course the finite branch-points may be located at different points.

With certain modifications, the method just discussed can be extended to arbitrary *algebraic functions*. These are functions that are obtained by solving (theoretically, at least) an algebraic equation $f(z, w) = 0$ between z and w. The new feature in this situation is that we now have *two* Riemann surfaces to consider, one over the z-plane for the function $w = w(z)$ and one over the w-plane for the function $z = z(w)$. The first of these surfaces was identical with the z-plane itself in all the simple special cases of *rational* functions that we have discussed so far; for, $w(z)$ is a single-valued function of z if it is rational in z. In the general algebraic case (as also in our special cases), the two Riemann surfaces are mapped onto each other one-to-one and conformally by the two functions $w(z)$ and $z(w)$. Thus, for example, $w^n = z^m$ (with relatively prime integers n, m) maps a surface of m sheets over the w-plane, with branch-points at 0 and ∞, onto a surface of n sheets over the z-plane, with branch points at 0 and ∞.

We shall conclude this section with an *example*. The Riemann surface of the function

$$w = \sqrt{(z-a)(z-b)(z-c)(z-d)}$$

has its branch-points at $z = a, b, c, d$; for at $z = a$, for instance, w can not be developed according to integral powers of $z = a$. The surface over the z-plane has two sheets. It is obtained by joining the two sheets along two arcs each of which connects two of the branch-points (one of them, say, a and b, the other one, c and d) without meeting the other branch-points. This surface is mapped by $w = w(z)$ onto a Riemann surface of four sheets over the w-plane. It is not possible to map the first Riemann surface one-to-one onto a simple plane by any function whatsoever if a, b, c, and d are four distinct points; for if we draw in one of the sheets a curve whose projection into the z-plane loops around a and b while leaving c and d in its exterior, then this curve does not decompose the surface into two separate regions, since it is possible to connect the two sides of the given curve by means of a suitably chosen other curve that does not cross the given curve (cf. Fig. 12).

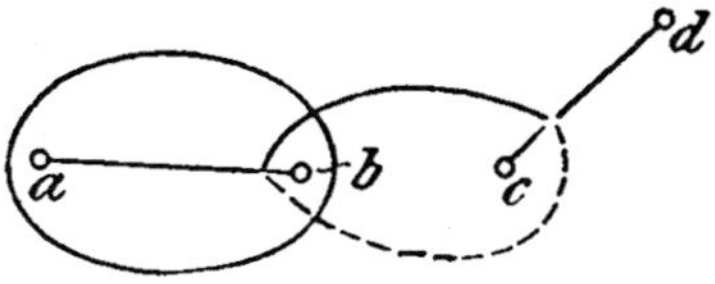

FIG. 12

But if our Riemann surface could be mapped one-to-one onto the z-plane, then there would have to be closed curves in the z-plane, according to the above, that do not decompose the plane. There are no such curves; we shall not prove this fact here but merely appeal to the reader's geometric intuition. We only wish, in passing, to call the reader's attention to certain facts that are of fundamental importance for various deeper problems of Function Theory, e.g. for the problem of uniformization upon which we shall touch a few more times. This problem, as we saw on pp. 61-62, depends on the conformal mapping of the Riemann surface onto a *simple* region of the plane. For the Riemann surfaces to which we were led by considering rational functions $w = f(z)$ and their inverses $z = \varphi(w)$, such a mapping was always possible; but we can easily guess here that such a one-to-one mapping will not always be possible in more general cases. The Riemann surfaces of the rational functions that we considered were mapped one-to-one onto the simple z-plane by their very make-up in terms of the function $z = \varphi(w)$. In this case, every function of w that was single-valued on the Riemann surface could be regarded as a single-valued function of z.

CHAPTER THREE

General Considerations

§ 9. The Relation Between the Conformal Mapping of the Boundary and that of the Interior of a Region

The following theorem has already been used implicitly several times, although it has not so far been necessary, in view of the simplicity of our examples, to state it explicitly:

THEOREM. *Let there be given a simple and simply connected region R lying entirely in the finite part of the z-plane and having one single boundary curve $\mathfrak{C}$. Let $\mathfrak{C}$ consist of a finite number of arcs of curves that are continuous and differentiable. Let the function $w = f(z)$ be regular and let $|f(z)|$ be less than some fixed finite bound within and on the boundary of R. Let the boundary curve $\mathfrak{C}$ of R be mapped one-to-one by $w = f(z)$ onto a closed curve $\mathfrak{C}'$ that does not intersect itself. Then the function $w = f(z)$ maps the region R one-to-one onto a finite, simple region R' whose boundary is $\mathfrak{C}'$.*

Proof. To make things easier, we assume familiarity with the fact that the curve $\mathfrak{C}'$ divides the

w-plane into exactly two regions, a finite one called the *interior* of $\mathfrak{C}'$, and an infinite one, the *exterior*. It is then obvious (by the Preservation-of-Neighborhoods Theorem) that the image region R' must have a region in common with one of the two regions just mentioned, and that the entire boundary of R' is given by the curve $\mathfrak{C}'$. But then R' can not intersect the exterior of $\mathfrak{C}'$, since otherwise R', being a finite region, would need some boundary curve *outside* of $\mathfrak{C}'$ to separate it from ∞. Therefore R' must have a region in common with the interior of $\mathfrak{C}'$. Let W be any point of this interior; then we must show that R' covers W exactly once.

Let us apply Cauchy's Integral Formula to the function $w = 1$ and the interior of $\mathfrak{C}'$; we obtain

$$1 = \frac{1}{2\pi i} \int_{\mathfrak{C}'} \frac{dw}{w - W}.$$

If we now make the substitution $w = f(z)$, which is one-to-one on the curve $\mathfrak{C}'$, we find that

$$1 = \frac{1}{2\pi i} \int_{\mathfrak{C}} \frac{f'(z)dz}{f(z) - W}.$$

Here we must take the integral over the boundary $\mathfrak{C}$ of R in such a way that the interior remains to the left as we traverse the boundary. But as is well-known, the last equality means that $f(z)$ takes on the value W exactly once in R. (If the path of integration were traversed in the opposite direction, the meaning would be that $f(z)$ has a pole in R, contrary to our assumption.) Hence the image R' of R covers every point of the interior of $\mathfrak{C}'$ exactly once, and since R' has no points in the exterior of $\mathfrak{C}'$ (as we saw a few lines back), it follows that R' is identical with the interior of $\mathfrak{C}'$; our proof is complete.

The theorem just proved allows us to make an inference from the mapping of the boundary of the region to the mapping of the whole region itself. This theorem can easily be extended to cover also the case where one or the other of the curves $\mathfrak{C}$, $\mathfrak{C}'$ passes through the point at infinity, provided only that the function $f(z)$ has no pole within the region R itself. This last condition, however, is essential to the validity of the theorem, as we shall soon see by means of examples—for instance in § 12, where we shall study the function $w = z + 1/z$.

§ 10. Schwarz' Principle of Reflection

We shall frequently have to make use of another theorem, of no less importance than the preceding one, called Schwarz' *Principle of Reflection.* This theorem will enable us, under certain conditions, to draw conclusions as to the behavior of a mapping in some region if we know how the mapping works in another region. The simplest such case is that of a conformal mapping of a region under which a part of a straight line goes into part of a straight line, the mapping being regular at every point of the first-mentioned line.

We shall first deal with the case where *both lines are part of the real axis*. From this fact we can deduce easily that any two points symmetric with respect to the real axis must be mapped onto another such pair. For let a be a point of a segment of the real axis on which $w = f(z) = \mathfrak{P}(z - a)$ assumes real values; then the coefficients in $\mathfrak{P}(z - a)$ must be real, since they are, to within real factors, the derivatives of $f(z)$ at $z = a$, and these derivatives are real on the segment in question, just as is $f(z)$ itself. It follows that the values which $f(z)$ takes on at two complex conjugate values of z must themselves be complex conjugates. This holds, to start with, for points within the circle of convergence of the power series. By analytic continuation along paths symmetric with respect to the real axis, we obtain pairs of power series whose respective coefficients are complex conjugates, so that our theorem holds for every region that can be reached by continuing the given function analytically from the initial expansion at $z = a$.

Having treated this special case, we can now pass to the somewhat more general case in which the mapping takes a part of *any* straight line L_z

into a part of *any other* straight line L_w. We select linear functions $z = \alpha z_1 + \beta$ and $w = a w_1 + b$, representing rigid motions and which are such that real values of z_1 and w_1 correspond to values of z and w on the given lines L_z and L_w, respectively. Points symmetric with respect to the real z_1-axis (or w_1-axis) correspond via these auxiliary functions to points symmetric with respect to the line L_z (or L_w, respectively). Now let $w = f(z)$ be the given function that connects z and w. Then z_1 and w_1 are connected by the relation $w_1 = \dfrac{f(\alpha z_1 + \beta) - b}{a}$, which maps two parts of the real axis onto each other. Hence it takes complex conjugate values of z_1 into complex conjugate values of w_1. Therefore if

$$w_1 = \frac{f(\alpha z_1 + \beta) - b}{a},$$

then we also have $\bar{w}_1 = \dfrac{f(\alpha \bar{z}_1 + \beta) - b}{a}$. Now if we substitute back: $z = \alpha z_1 + \beta$, $w = a w_1 + b$, it follows that the given mapping $w = f(z)$ takes points symmetric with respect to L_z into points symmetric with respect to L_w, as we wished to prove.

The same type of argument as has just been used to extend our initial statement to the case of two arbitrary lines, can also be applied to generalize the statement still further, namely to the case in which it is any two *circular arcs* that correspond to each other under a given mapping. Let us recall that by reflection in a circle we mean mapping by reciprocal radii (or inversion) (cf. p. 17), and that a linear mapping that takes a straight line into a circle takes points related by reflection in the line into points related by reflection in the circle (cf. p. 39). We are thus led to the following general theorem **(Schwarz' Reflection Principle):**

If $w = f(z)$ *is analytic on a circular arc* k *of the* z*-plane and if the values it assumes on* k *lie on a circular arc* K *of the* w*-plane, then the values assumed by* $f(z)$ *on points related by reflection in* k *are in turn related by reflection in* K.

The hypothesis that $f(z)$ be analytic on the arc k can be relaxed; it is sufficient to assume that $f(z)$ is *analytic in the interior of a region of whose boundary the arc* k *forms a part, and that* $f(z)$ *is continuous in the extended domain consisting of the region and the arc* k. The analyticity of $f(z)$ on k itself is a consequence of this assumption. Because of the above-mentioned theorem of p. 39, it suffices to prove the last statement in the case that k is part of the real axis, that $f(z)$ assumes real values on k, that $f(z)$ is analytic within, and on the curved part of the periphery of, a semicircular region with center on k, and that $f(z)$

is continuous in the domain consisting of the interior points of the semi-circular region together with the points of k. The proof then rests on the fact that for values of z from the interior of the semi-circle, $f(z)$ can be represented by means of the Cauchy Integral Formula, the path of integration being the periphery of the semi-circular region; this fact can be established by a limiting process in which we first replace k by a neighboring chord parallel to k. Now if we consider the semi-circular region obtained from the above by reflection in k, and if we define in this mirror-image an analytic function $f_1(z)$ by setting $f_1(z) = \overline{f(\bar{z})}$, then we see that $f_1(z)$ coincides on k with $f(z)$. If z is now an interior point of the *new* region, and if we integrate over its periphery in the positive sense, we obtain

$$f_1(z) = \frac{1}{2\pi i}\int \frac{f_1(\zeta)}{\zeta - z}\, d\zeta .$$

If on the right-hand side of this formula z is chosen from *outside* the new semi-circle, for instance from the interior of the original semi-circular region, then by the Cauchy Integral Theorem, the left-hand side must be zero. Corresponding statements hold for the original semi-circle if in the expression

$$\frac{1}{2\pi i}\int \frac{f(\zeta)}{\zeta - z}\, d\zeta,$$

where the path of integration is the positive periphery of the original region, we choose for z a value from the interior of the new semi-circle. If we add the last two expressions, we obtain

$$\frac{1}{2\pi i}\int\frac{f(\zeta)}{\zeta - z}\,d\zeta + \frac{1}{2\pi i}\int\frac{f_1(\zeta)}{\zeta - z}\,d\zeta.$$

The first of these two integrals is to be taken over the original periphery and the second over the new one, both in the positive sense. The sum of the two integrals represents $f(z)$ if z is in the interior of the original semi-circle, and represents $f_1(z)$ if z is in the interior of the new semi-circle. But since $f_1(z) = f(z)$ holds on k, and since the two integrations along k are in opposite directions, the above sum of integrals may be written as a single integral taken over the periphery of a full circle. But it then represents a function that is regular everywhere in the interior of this circle, that coincides with $f(z)$ within the original and with $f_1(z)$ within the new semi-circle, and that must therefore, because of our continuity assumption, coincide with both $f(z)$ and $f_1(z)$ on k. Hence we have proved that $f(z)$ is also analytic at all points of k.

CHAPTER FOUR

Further Study of Mappings Represented by Given Functions

§ 11. Further Study of the Geometry of $w = z^2$

As we saw in § 7, the function $w = z^2$ maps the z-plane onto a Riemann surface of two sheets over the w-plane, with branch-points at 0 and ∞. In order to study this mapping in greater detail, we shall separate the real and imaginary parts. This method can be applied successfully to the investigation of other functions as well. Setting $w = u + iv$ and $z = x + iy$, we find that $u = x^2 - y^2$, $v = 2xy$.

The straight lines $u = \text{const.}$ correspond to the equilateral hyperbolas $c = x^2 - y^2$, and the straight lines $v = \text{const.}$ correspond to the equilateral hyperbolas $c = 2xy$, which are orthogonal to the first-named hyperbolas. The two branches of a hyperbola correspond to the two straight lines that are superimposed on each other on the two sheets of the Riemann surface; for, the branch-point $w = \infty$ separates these two lines just as $z = \infty$ separates the two branches of the hyperbola. The two lines differ only in that the

z-values to which they correspond are negatives of each other; the values of u and v as such remain unchanged if both x and y are replaced by their negatives. Fig. 13 shows the hyperbolas $u = c$ both for positive and for negative values of c. A region of the z-plane bounded by a branch of such a hyperbola and not containing $z = 0$, is mapped by the function $w = z^2$ onto a half-plane bounded by a straight line in the w-plane and not containing $w = 0$. The mapping of the first region onto the second (i.e. onto the half-plane) is one-to-one and isogonal. The z-region bounded by the other branch of the same hyperbola is mapped onto the same half-plane in the other sheet of the Riemann surface over the w-plane. We have thus found out how to map the "interior" of an equilateral hyperbola onto a half-plane, and since we already know how to map a half-plane onto the interior of a circle, we can also map the interior of an **equilateral hyperbola** conformally onto the interior of a circle.

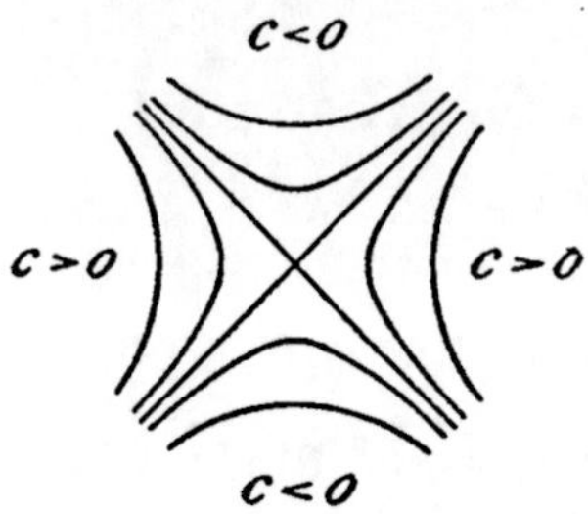

FIG. 13

The equations $u = x^2 - y^2$, $v = 2xy$, which embody the mapping $w = z^2$, can also be used to find the images under this mapping of the straight lines $x =$ const. and $y =$ const. From $x = c$ it follows that $y = v/2c$, hence that $u = c^2 - v^2/4c^2$, or $v^2 = 4c^2(c^2 - u)$. This is the equation of a parabola whose axis is $v = 0$ and whose focus is at $u = 0$, $v = 0$, and which opens toward $u < 0$. Similarly, the straight lines $y = c$ are mapped onto the parabolas $v^2 = 4c^2(u + c^2)$. They also have the line $v = 0$ as their axis and the point $u = 0$, $v = 0$ as their focus, but they open toward $u > 0$.

The exterior of such a **parabola** (shaded in Fig. 14) of the w-plane is mapped by the function $w = z^2$ conformally onto a half-plane not containing $z = 0$; the mapping of the interior is not isogonal at $z = 0$. Let us consider what happens under the mapping to just the upper half of the interior of the parabola,

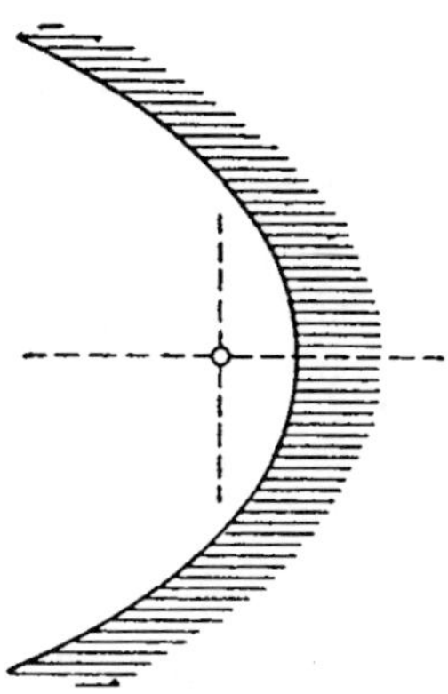

FIG. 14

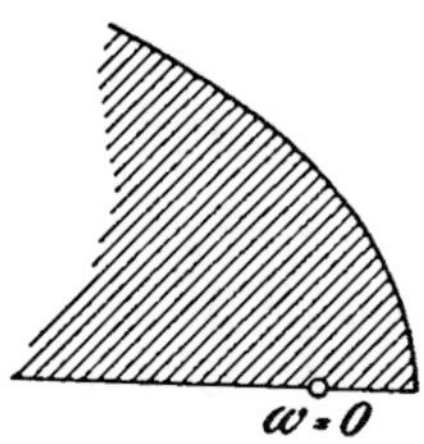

FIG. 15

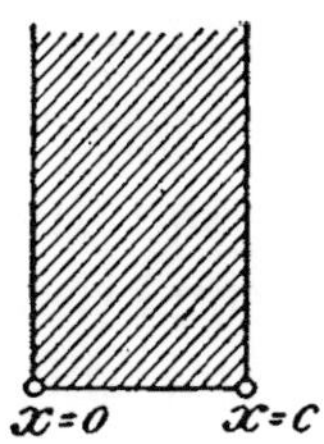

FIG. 16

i.e. to that half where $v > 0$, shown in Fig. 15. The pre-image of the negative real axis $u < 0$ is the positive and the negative imaginary axis of the z-plane; let us choose that branch of the inverse function which gives the positive imaginary axis. If we then traverse $u < 0$ from zero to infinity, the corresponding point in the z-plane will traverse the positive imaginary axis from zero to infinity. Therefore the pre-image of the upper half of the interior of our parabola must be to the right $(x > 0)$ of the imaginary axis. Thus the segment from $w = 0$ to $w = c^2$ (the vertex of the parabola) of the positive real axis goes into the segment from $z = 0$ to $z = |c|$ of the positive real axis of the z-plane. The arc of the parabola itself goes into the upper half $(y > 0)$ of the line $x = |c|$. Thus the upper half of the interior of the parabola, shaded in Fig. 15, has as its pre-image under the mapping (more precisely, under the chosen branch of the inverse mapping) the half-strip shaded in Fig. 16.

What happens to the other half of the interior of the parabola under our mapping? We observe that the half-line $v = 0$, $u < 0$ goes into the imaginary axis (exclusive of $z = 0$) of the z-plane, and that the parabola is symmetric with respect to this half-line; therefore by Schwarz' Reflection Principle, we obtain the image (or rather, pre-image) of the other half of the interior of the parabola by reflecting the half-strip of Fig. 16 in the half of the imaginary axis on which it abuts. The whole interior of the parabola is thus seen to be mapped onto a vertical half-strip double the width of each of the two just considered, lying above the real axis of the z-plane and extending from $x = -|c|$ to $x = +|c|$; the vertical half-lines bounding this double half-strip correspond to the parabola itself, and the segment from $w = 0$ to $w = c^2$ of the real axis of the w-plane corresponds to the line-segment from $x = -|c|$, $y = 0$ to $x = +|c|$, $y = 0$. If we wish to deal with only one of the two branches of the function $z = \sqrt{w}$, we must think of the axis of the parabola as being cut from the vertex to the focus. Instead of the branch that we have just been using in the above description, we could use the other branch; then the parabola, with the same cut as before, would be mapped onto the mirror image below the real axis of the z-plane of our above double half-strip. Finally, let us go back to the mapping of the region shaded in Fig. 15 onto that shaded in Fig. 16; suppose that this time we cut the axis of the parabola from $w = 0$ to ∞ along $u < 0$, instead of as before. Noting that the segment from $w = 0$ to $w = c^2$ goes into the segment $0 < x < |c|$ of the real axis of the z-plane, we then see, by the same sort of reasoning as before, that the whole interior of the parabola is mapped by one branch of $\sqrt{w}$ onto the strip $0 < x < |c|$, and is mapped onto the strip $-|c| < x < 0$ by the other branch of $\sqrt{w}$.

We shall come back to this discussion in § 15, No. 10.

The function $w = z^2$ maps the circle

$$(x - r)^2 + y^2 = r^2, \ r > 0,$$

onto the cardioid

$$(u^2 + v^2)^2 - 4r^2u(u^2 + v^2) - 4r^4v^2 = 0 .$$

This cardioid, shown in Fig. 17, is the epicycloid obtained when a circle of radius r^2 rolls on the dotted circle of Fig. 17, also of radius r^2. Under our mapping, the interior of the circle $|z - r| < r$ is mapped one-to-one onto the interior of the cardioid. Incidentally, the cardioid is the image under the mapping $w = 1/z$ of the parabola $4r^4y^2 + 4r^2x - 1 = 0$; this mapping takes the exterior region of the parabola onto the interior of the cardioid.

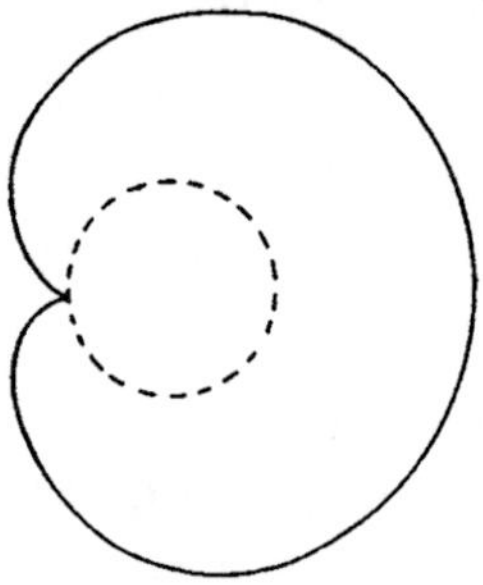

FIG. 17

§ 12. $w = z + 1/z$

To any given value of w, there correspond two values of z. Therefore the z-plane is mapped onto a Riemann surface of two sheets over the w-plane by the function $w = z + 1/z$. The branch-points of this surface are at $w = \pm 2$. They correspond to the points $z = \pm 1$.

A detailed study of the mapping will be simplified by setting $w = u + iv$, $z = \varrho e^{i\varphi}$ and then separating real and imaginary parts. This yields $u = (r + 1/r) \cos\varphi$, $v = (r - 1/r) \sin\varphi$. The circles $r = \text{const.}$ of the z-plane are mapped onto ellipses with semi-axes $r + 1/r$, $|r - 1/r|$. For every $r \neq 1$, the two circles whose radii are r and $1/r$ are mapped onto the same ellipse of the w-plane, corresponding to the two sheets of the Riemann surface. The circle $r = 1$ is mapped onto the line-segment from $w = -2$ to $w = +2$, i.e. onto the segment joining the common foci of all the above ellipses.

The straight lines $\varphi = \text{const.}$ are mapped onto the hyperbolas $\frac{u^2}{\cos^2\varphi} - \frac{v^2}{\sin^2\varphi} = 4$, whose semi-axes are $2|\cos\varphi|$, $2|\sin\varphi|$. The foci of these hyperbolas are the same as those of the above ellipses, viz. $w = -2$ and $w = +2$. The two rays (half-lines) $\varphi = 0$ and $\varphi = \pi$ correspond to the part of the real axis that connects $w = +2$ *via* $w = \infty$ to $w = -2$. Any other pair of rays $\varphi = a$ and $\varphi = -a$ is mapped onto the two branches of the same hyperbola. The ellipses and hyperbolas with which we are dealing constitute a system of confocal conics.

The inverse of our function maps the region bounded by the two branches of one and the same hyperbola—a region not containing any branch-points—onto a region of the z-plane bounded by two half-lines that join 0 and ∞, viz. the half-lines $\varphi = a$ and $\varphi = \pi - a$. A region that is bounded by two branches of two different hyperbolas on the Riemann surface is mapped onto a sector of the z-plane bounded by two half-lines that join 0 and ∞. It is just as easy to determine what happens to the exterior of one of the above ellipses under the mapping. Such an exterior region is free of branch-points, and is mapped either onto the interior or onto the exterior of the corresponding circle in the z-plane, depending on the sheet of the Riemann surface to which the region belongs. As an important special case, we note that the full w-plane, cut along the real axis from $w = -2$ to $w = +2$, is mapped onto the interior or the exterior of the circle $|z| = 1$.

There is no comparable simplicity in the result to which we are led by considering the effect of the mapping on the *interior* of our conics. Here, too, we shall find, just as in the preceding section, that the branch-points create a certain amount of trouble. Let us first consider those halves of our conics that are located above the real axis, and let us start with an ellipse. A **semi-ellipse** (Fig. 18) is mapped onto a quadrangular region whose sides are arcs of circles; the sides are made up of two concentric semi-circles above the real axis of the z-plane and the two segments of the real axis that join the semi-circles (cf. Figs. 18 and 19, where corresponding points are labelled with the same letters). One of the semi-circles is part of the unit circle, being the image of the segment from

FIG. 18

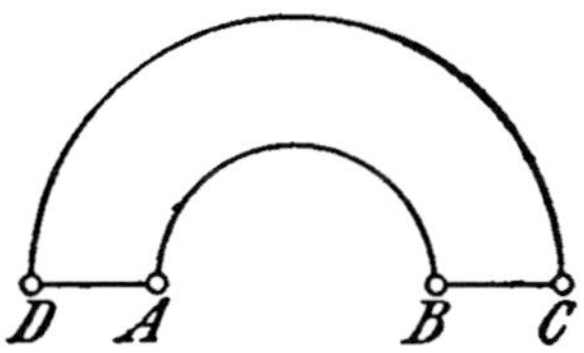

FIG. 19

$w = -2$ to $w = +2$ of the real axis of the w-plane; the diameter of the other semi-circle is greater than or less than that of the unit circle, depending on the sheet of the Riemann surface in which the ellipse is considered as being located, and depending also on the branch of the inverse function that is used to accomplish the mapping. The semi-axes of the ellipse being $r + 1/r$ and $|r - 1/r|$, the radii of the circles will be 1 and r.

In the case of the **hyperbola,** the situation is similar. Its upper half is mapped onto a triangular region bounded by a part of the real axis, a part of another straight line that would pass through $z = 0$ if continued, and an arc of the unit circle (cf. Figs. 20 and 21).

For both of the above cases, the way to determine the image of the other (lower) half is to reflect the image of the upper half in the unit circle or in the real axis of the z-plane, in accordance with Schwarz' Reflection Principle. In this

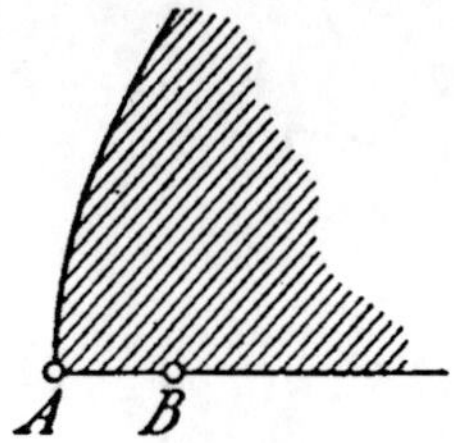

FIG. 20

FIG. 21

way, by reflection in the unit circle we obtain as the image of the interior of the ellipse, cut from the vertices to the foci, the quadrangular region shown in Fig. 23. (In Figs. 22 and 23, corresponding points are labelled with the same letters.)

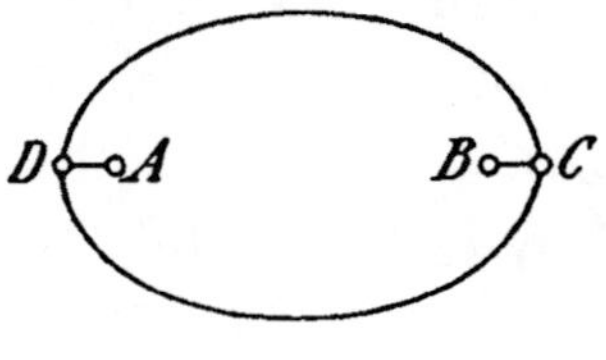

FIG. 22

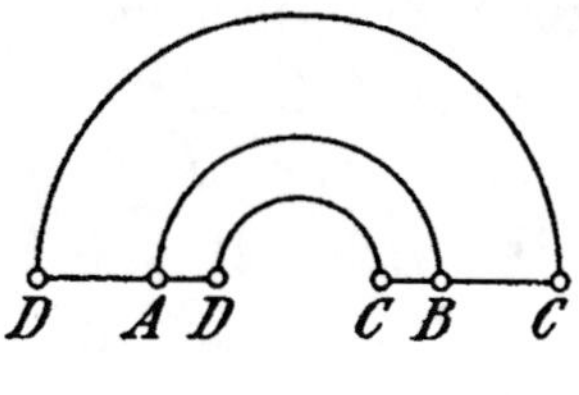

FIG. 23

If we cut instead along the major axis of the ellipse between the foci, we must reflect the image of the upper half in the real axis, obtaining a circular annulus (cf. Figs. 24 and 25).

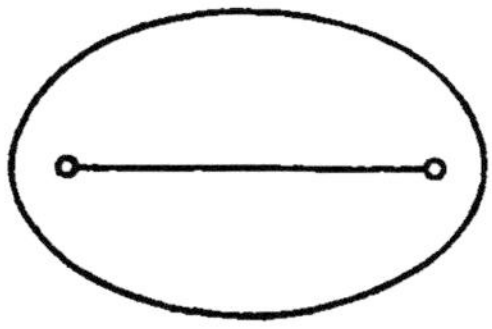

FIG. 24

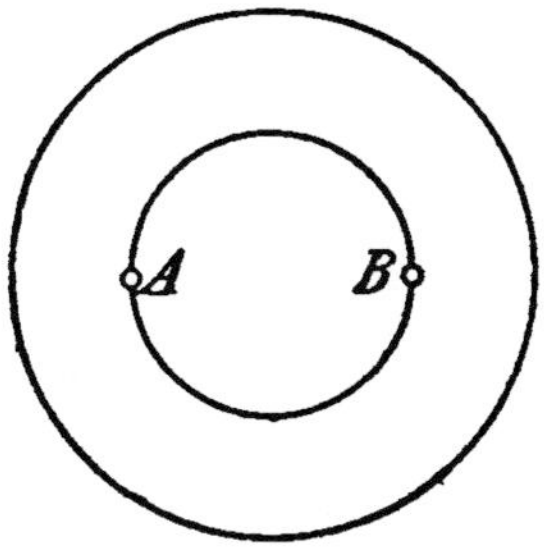

FIG. 25

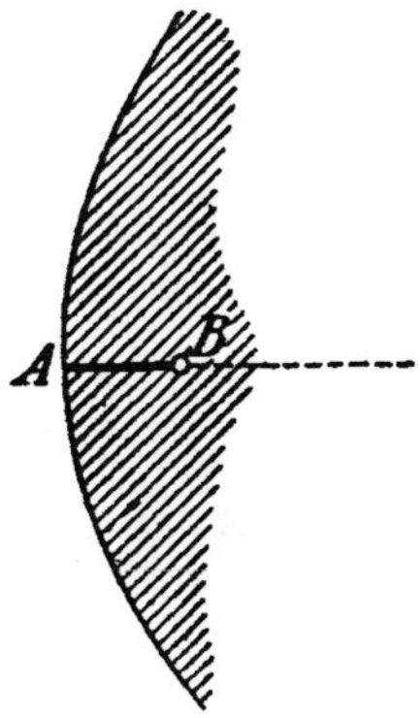

FIG. 26

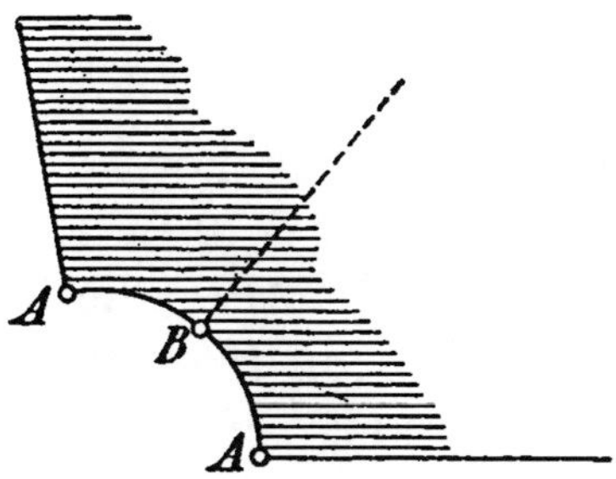

FIG. 27

We can proceed similarly in the case of the hyperbola and obtain the two pairs of diagrams 26, 27 and 28, 29.

The function $y = z^{\frac{\pi}{\delta}}$, where δ denotes the angle at D of the sector of Fig. 29, maps this sector onto a half-plane.

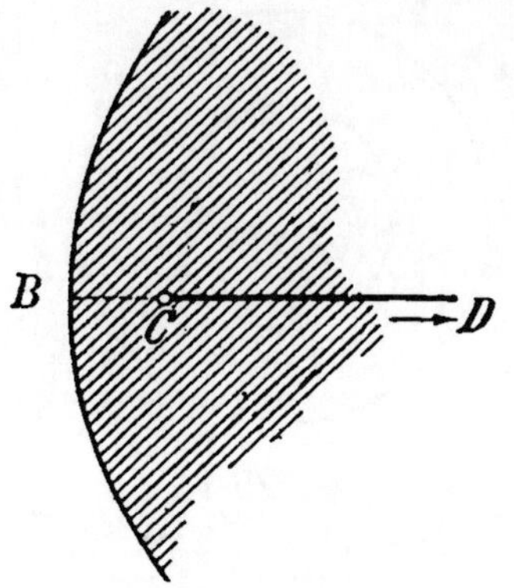

FIG. 28

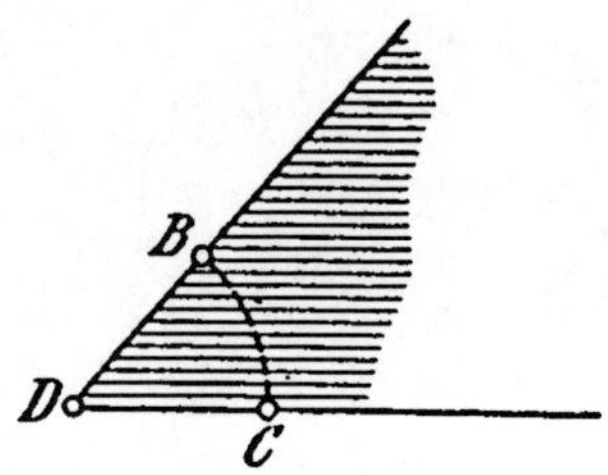

FIG. 29

Let us finally determine the image under our mapping $w = z + 1/z$ of a circle K passing through $z = -1$. The mapping fails to be isogonal at $z = 1$ and at $z = -1$, and we must therefore specify whether we want $z = 1$ to be inside K or on K or outside K. We can make use of our study of the function $w_1 = \mathfrak{z}^2$, by first setting $\mathfrak{z} = (z + 1)/(z - 1)$ and noting that the mapping $w = z + 1/z$ is then expressed by $w = 2(w_1 + 1)/(w_1 - 1)$. Observe that $\mathfrak{z} = (z + 1)/(z - 1)$ maps K onto a circle through $\mathfrak{z} = 0$, which is mapped by $w_1 = \mathfrak{z}^2$ onto a cardioid in the w_1-plane like that in Fig. 17. To obtain the final image of K in the w-plane, we must therefore subject this cardioid to the mapping $w = 2(w_1 + 1)/(w_1 - 1)$. The shape of the image curve can also be obtained directly, by adding the vectors z and $1/z$ as z is made to traverse the circle K. The result is shown in Fig. 30; here, K' is the circle onto which K is mapped by $1/z$, and E is the unit circle. This diagram applies to the case that the point $z = 1$ is outside K; in this case our

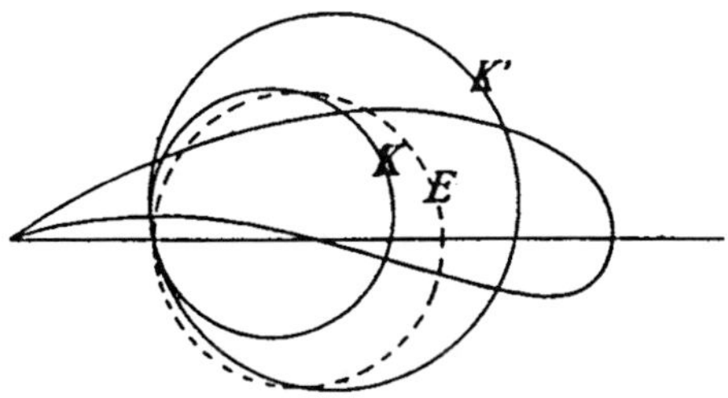

FIG. 30

function maps the interior of the circle onto the simple interior of the image curve in Fig. 30. If $z = 1$ is an interior point of K, then the exterior of K is mapped onto the simple exterior of a similar image curve; for in this case, $\mathfrak{z} = (z+1)/(z-1)$ maps the exterior of K onto the interior of a circle through $\mathfrak{z} = 0$ that must itself contain the point $\mathfrak{z} = 1$, as the image of $z = \infty$. The interior of the latter circle is mapped by $w_1 = \mathfrak{z}^2$ onto the interior of a cardioid that contains $w_1 = 1$. Since $w = 2(w_1 + 1)/(w_1 - 1)$ sends the point $w_1 = 1$ to $w = \infty$, and since the periphery of the image of K will again be of the same shape as in Fig. 30, it follows that the simple exterior of the image curve is the image of the exterior of K, as stated above.

The image curve shown in Fig. 30 plays a role in aerodynamics, under the name "Jukowski profile." It represents one of the possibilities for the shape of the cross-section of an air-foil (airplane wing). The mapping just described, of the exterior of this profile onto the exterior of a circle, reduces the problem of determining the stream-lines of an incompressible medium around the profile to the problem of determining the stream-lines about a circular profile. This is because the stream-lines in question are represented in terms of potential functions, and as is well known, a potential function is transformed by any conformal mapping into another potential function, such functions being the real parts of analytic functions. See also § 20 below.

§ 13. The Exponential Function and the Trigonometric Functions

The function $w = e^{iz}$ is periodic of period 2π, since $e^{iz+2i\pi} = e^{iz}$. Therefore all the values which this function assumes anywhere are taken on in some part of the plane, namely in the strip bounded by the lines $x = 0$ and $x = 2\pi$ of the z-plane $(z = x + iy)$. To see how the function maps this strip into the w-plane, we set

$$w = re^{i\varphi},\ z = x + iy$$

and therefore have $r = e^{-y}, \varphi = x$. This shows that the straight lines $x =$ const. are mapped onto the rays (half-lines) $\varphi =$ const. of the

w-plane, while the lines $y = \text{const.}$ are mapped onto the circles $r = \text{const.}$, so that the whole w-plane is covered just once by the image region of the strip. The boundary lines of the strip are each mapped onto the positive real axis of the w-plane. If we think of the whole z-plane as being divided up into congruent vertical strips each of width 2π, one of which is the strip considered above, we see that the mapping of each strip gives an image covering the whole of the w-plane. Thus the image of the entire z-plane under the mapping $w = e^{iz}$ turns out to be a Riemann surface of infinitely many sheets, with its branch-points at $w = 0$ and $w = \infty$. The function $z = -i \log w$ maps this Riemann surface onto the simple z-plane. Therefore any function that is single-valued on this Riemann surface, i.e. any function of w that has no branch-points other than $w = 0$ and $w = \infty$ (at which two points it may have branch points of any order whatsoever), can be regarded as a single-valued function of z; we mention this with a view toward the developments in a later section.

Let us apply our mapping to the quadrangular region of Fig. 23 onto which we mapped the interior of an ellipse, with suitable cuts, in the preceding section. We think of this quadrangular region as being located in the $w = e^{iz}$-plane, and we shall map it into the z-plane by means of the function $z = -i \log w$. Its image is a rectangle; the image of the boundary line along the positive real axis is a part of the straight line $x = 2\pi$ (provided that we choose the appropriate branch of the logarithm), and the images of the two circular arcs are parts of the lines $y = 0$ and $y = \log(1/r)$ respectively, since the radii of the circles are 1 and r. Finally, the image of the boundary line along the negative real axis is the straight line $x = 0$.

The above considerations enable us to obtain an insight also into the mappings represented by the **trigonometric functions.** We recall that

$$\sin z = \frac{e^{iz} - e^{-iz}}{2i}, \quad \cos z = \frac{e^{iz} + e^{-iz}}{2}, \quad \tan z = \frac{e^{iz} - e^{-iz}}{(e^{iz} + e^{-iz})\, i}.$$

From this it is evident that we need merely combine suitably such mappings as have already been discussed. Let us first take up the mapping $w = \cos z$. We start by mapping the z-plane onto a $\mathfrak{z}$-plane by means of $\mathfrak{z} = e^{iz}$; we next pass to a $\mathfrak{w}$-plane by means of $\mathfrak{w} = \mathfrak{z} + 1/\mathfrak{z}$, and thence to the w-plane by means of $w = \mathfrak{w}/2$. This combination yields $w = \cos z$. Therefore the mapping $w = \cos z$ takes the straight lines and circles of the z-plane that we considered above in connection with $w = e^{iz}$, into the confocal ellipses and hyperbolas with their foci at $w = \pm 1$.

The mapping $w = \sin z$ is now taken care of by the remark that $\sin z = \cos(\pi/2 - z)$, and we are in full control of $w = \tan z$ by setting

$$\mathfrak{z} = 2z, Z = e^{i\mathfrak{z}}, \ \mathfrak{w} = \frac{Z-1}{Z+1}, \ w = -i\mathfrak{w}.$$

Under $w = \tan z$, the parallels to the imaginary axis of the z-plane are mapped onto the circles passing through $w = +1$, and the parallels to the real axis of the z-plane are mapped onto the family of circles orthogonal to the ones just mentioned.

§ 14. The Elliptic Integral of the First Kind,

$$w = \int^{z} \frac{dz}{\sqrt{(z-a_1)(z-a_2)(z-a_3)(z-a_4)}}.$$

We have discussed earlier in this book (cf. p. 46) the Riemann surface of the square-root function appearing in the above integral. The surface has two sheets over the z-plane, and its branch-points are at a_1, a_2, a_3, a_4. Let us investigate the mapping to which the integral subjects the Riemann surface. We shall restrict ourselves here to the case of four *real* branch-points. It is then easy to see that the integral can be developed in powers (with non-negative whole exponents) of the local uniformizing variable corresponding to the place with whose neighborhood on the surface we happen to be concerned. This variable is $\sqrt{z-\alpha}$ if the place is a branch-point; it is $z - a$

if the place is an ordinary point of the surface; and it is $1/z$ if the place is at $z = \infty$, unless $z = \infty$ is a branch-point (so that the radicand is a polynomial of the third degree, instead of the fourth, a possibility to be subsumed under the heading of this section), in which case the variable is $1/\sqrt{z}$. To verify all this, we first write down the power-series expansion of the radical

$$W = \sqrt{(z-a_1)(z-a_2)(z-a_3)(z-a_4)}$$

which at $z = a_1$, for instance, is as follows:

$$W = \sqrt{z-a_1} \cdot \mathfrak{P}(z-a_1).$$

We must then integrate term by term, and obtain

$$w = \sqrt{z-a_1} \cdot \mathfrak{P}_1(z-a_1).$$

Here, as usual, $\mathfrak{P}$ and $\mathfrak{P}_1$ denote power series of the form $c_0 + c_1(z-a_1) + \cdots$. We see at the same time that the function represented by the integral is finite everywhere on the Riemann surface.

In spite of its *local* single-valuedness, we shall see shortly that our integral is not a single-valued function on the entire surface. To get a picture of the mapping which it represents, we first cut our Riemann surface along the real axis (recall that we assumed the a_i to be real), through both sheets, into four half-planes. Let us first deal with the mapping of one of these half-planes, say of an upper one; the mapping of the remaining ones can then be treated by Schwarz' Reflection Principle. For the sake of definiteness, we must first specify which of the values of the radical we want to use in the mapping of the half-plane that was chosen. Let $a_1 < a_2 < a_3 < a_4$ and set

$$z - a_k = r_k e^{i\varphi_k} \quad (k = 1, 2, 3, 4)$$

the angles φ_k being between 0 and π (cf. Fig. 31). Then for the mapping of our half-plane, we shall use

$$\sqrt{z - a_k} = r_k^{1/2} e^{i\frac{\varphi_k}{2}}.$$

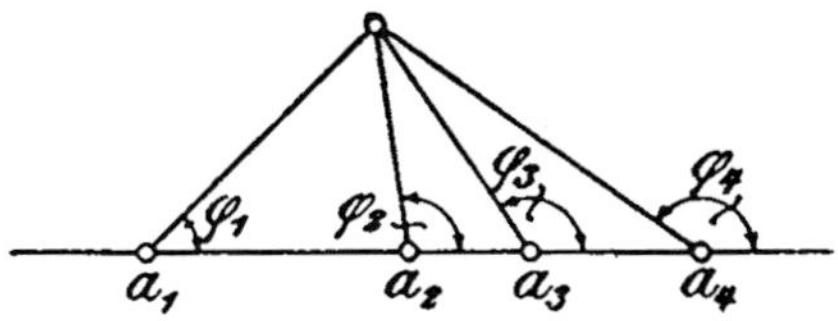

FIG. 31

This being agreed on, we see that the integrand is a positive pure imaginary for z between a_1 and a_2, that it is negative and real for z between a_2 and a_3, that it is a negative pure imaginary for z between a_3 and a_4, and finally that it is positive and real for z between a_4 and a_1. Considering now the integral itself, let us choose its lower limit to be at the point a_4. Then if z goes from a_4 *via* ∞ to a_1, along the real axis, then by what has just been observed, the integral w is always positive and increasing, so that w travels in the positive direction along the real axis of the w-plane, from 0 to some point ω_1. As z continues from a_1 to a_2, the integral will change because of positive pure imaginary contributions, so that w will travel in a constant direction, parallel to the imaginary axis of the w-plane, from ω_1 to some point $\omega_1 + \omega_2$. As z continues further, toward a_3, the integral w will again move parallel to the real axis, but this time back toward the imaginary axis; and finally, as z travels from a_3 to a_4, the corresponding w will move parallel to the imaginary axis and toward the real axis. Now note that w must return to $w = 0$ as z finally returns to a_4, since the integral taken along the entire real axis of the z-plane must have the value 0, by the Cauchy Integral Theorem. Thus w must have described a rectangle whose interior remained to the left as w traversed its boundary, just as the upper half-plane remained to the left as z went through its corresponding journey along the real axis of the z-plane; and since, as we saw above, the integral is everywhere finite, it follows from the results of § 9 that the mapping represented by the integral *maps the upper half-plane over the z-plane one-to-one onto the interior of our rectangle. The sides of the rectangle are of lengths* $|\omega_1|$ *and* $|\omega_2|$.

If we reflect the upper half-plane in the part from a_4 to a_1 (*via* ∞) of the real axis, we obtain a lower half-plane whose image under our mapping is then obtained by reflecting the rectangle in the corresponding side, the one from 0 to ω_1. If we want to pass from the one full sheet just obtained to the whole Riemann surface, we need only reflect this sheet in the two banks of, say, the line-segment from a_3 to a_4. Across this line-segment we may pass from one sheet to the other; the two sheets are joined cross-wise along the segment. We therefore obtain the image of the entire Riemann surface by reflecting in the line-segment from $-\omega_2$ to ω_2 the double rectangle obtained just before by reflection in $0\omega_1$ of the original rectangle. Thus the image of the entire Riemann surface is a large rectangle whose vertices are

$$-\omega_1-\omega_2,\ \omega_1-\omega_2, \omega_1+\omega_2,\ -\omega_1+\omega_2,$$

built up from four congruent smaller rectangles each of which corresponds to one of the four half-planes of which the Riemann surface is made up. It can now be seen that our integral is not a single-valued function on the Riemann surface; for in order to complete the construction of the closed Riemann surface from its four half-planes, we have yet to join suitably the remaining edges of the cut along the real axis that correspond to the sides of the rectangle. Consider, say, the segment from a_2 to a_3 of the real axis; along this segment, we must simply close up a slit in each of the two sheets. But at the points that are thus

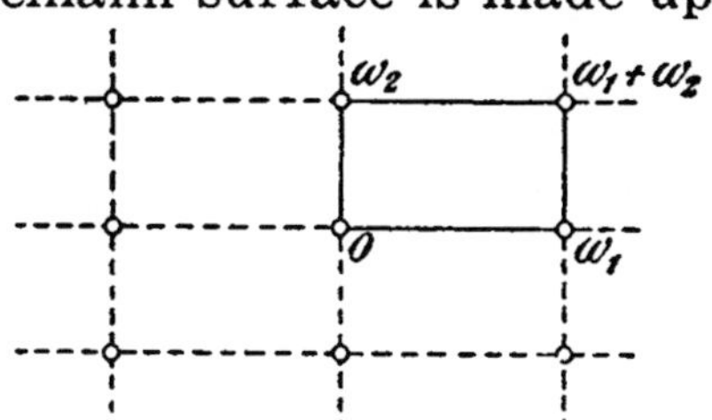

FIG. 32

brought into coincidence, and that up to now were on opposite banks of a cut, the integral evidently assumes values that differ by $2\omega_2$; for as can be seen from the way in which the large rectangle was obtained from the four small ones, any two points on the boundary of the large rectangle that differ by $2\omega_2$ correspond to one and the same point of the Riemann surface. The same holds for the other pair of parallel sides; any two points differing by $2\omega_1$ correspond to one and the same point of the Riemann surface. The joining of the cuts of the Riemann surface may be visualized in its image, the large rectangle, by thinking of each pair of parallel sides as being bent together and being joined so that points corresponding to the same point of the Riemann surface are made to coincide. To be sure, in carrying out this process of bending together and joining parallel sides, it is necessary to go out of the plane and to construct a closed surface in three-dimensional space. The result is the surface of a torus (or doughnut); for by joining one pair of parallel sides of the rectangle, we first obtain a sort of tubular hose, and by bending together and joining the two ends of the hose we obtain the torus. The Riemann surface under discussion is in this way mapped one-to-one and continuously onto a torus, which is thus suitable for the purpose of intuitive visualization of the structure of our Riemann surface.

Being essentially different from, say, the sphere, onto which we mapped the simple plane in a previous section (§ 3), the torus carries various closed curves that do not divide it into separate regions, just as does the Riemann surface. The fact that our integral, while regular on the whole surface, still fails to be a single-valued function on the surface is due exactly to the existence of such curves. Going back to Fig. 12 on p. 69, we observe that as z traverses the closed curve marked out in that diagram, the value of w changes by $2\omega_2$. This curve, however, does not decompose the surface, as was noted on p. 69.

It remains to study the overall distribution on the Riemann surface of the values of the integral of the first kind. This is a simple matter once we observe that our rectangle in the w-plane may be reflected in each of its sides, and that each such reflection has its counterpart in the reflection of the Riemann surface in the corresponding straight line. By continuing indefinitely with such reflections, we gradually obtain all the values that the integral can take on at a given place of the Riemann surface. This process of repeated reflections leads to a simple and complete covering of the entire finite w-plane by congruent rectangles, and shows that the totality of values that w can take on at a given place of the Riemann surface, is the totality of numbers of the form

$$w = w_1 + 2h\omega_1 + 2k\omega_2 \, ,$$

where w_1 is a specific one of these values while h and k are any rational integers whatever. Geometrically, we obtain this totality of points of the w-plane by subjecting the rectangle in which the initial w_1 lies, to the translations by $2h\omega_1 + 2k\omega_2$, each of which makes it coincide with some other rectangle of our tiling of the w-plane, and then marking in each such rectangle the point whose position corresponds to that of w_1 in the initial rectangle.

Instead of considering our integral as a function defined on a Riemann surface of two sheets, as we have done so far, we could construct a special Riemann surface for it which would be built up from an infinite number of duplicates of the above surface of two sheets in the same way as the w-plane is built up from the rectangles with which we just covered it; the w-plane would then be a one-to-one isogonal image of the new Riemann surface. This latter is also called a covering surface of the original Riemann surface of the radical.

Every function that is of rational character[1] in the local uniformizing variable everywhere on the Riemann surface of the radical, is obviously of

[1] A function is said to be of *rational character* in the neighborhood of a place a if it either is regular at a or has a pole at a.

rational character everywhere in the finite w-plane when considered as a function of w; for, at all places of the Riemann surface, w itself represents a local uniformizing variable, according to the definition we gave in a previous section (§ 7), since the integral of the first kind maps the neighborhood of any given point of the surface onto a simple region of the w-plane. Now a function that is of rational character everywhere in the finite w-plane must also be a single-valued[2] function of w. Therefore *any function that is single-valued everywhere on the Riemann surface of the radical becomes a single-valued function of w if the independent variable z is replaced in the function by its value in terms of w.* The Riemann surface of the integral of the first kind is simply the common Riemann surface of all functions of the kind just described. It can therefore be constructed without previous knowledge

[2] This follows from the Monodromy Theorem, which states that an analytic function, regular at all points of a simply-connected region except for poles, must be single-valued in this region. The proof of this theorem, in brief, is that if there were a closed curve in the region such that the function undergoes a change in its value upon one full traversal of the curve, then this change would still remain if the curve were shrunk to a point in the region, thus contradicting the assumption of the theorem. For the details, cf. Bieberbach, *Funktionentheorie,* Vol. 1 (Chelsea, 1945), or Carathéodory, *Theory of Functions,* Vol. 1 (Chelsea, 1953).

of the properties of the integral of the first kind. Every function that maps the surface onto a simple region can be considered, in place of the integral of the first kind, as a uniformizing function for the whole class of functions mentioned above. (We shall not be concerned here with the fact that all such mapping functions are linear functions of the integral of the first kind.)

According to the above discussion, it must be possible, in particular, to represent z and the radical W as single-valued functions of w. These functions turn out to have a further important property; they are **doubly-periodic.** For since the points $w_1 + 2h\omega_1 + 2k\omega_2$ all correspond to the same place on the Riemann surface as does the point w_1, it follows that $z(w)$ and $W(w)$ each takes on the same values at all these points. This means that the following functional equations hold:

$$z(w + 2h\omega_1 + 2k\omega_2) = z(w), \; W(w + 2h\omega_1 + 2k\omega_2) = W(w).$$

They express what is meant by calling $z(w)$ and $W(w)$ doubly-periodic functions.

In order to tie these matters up with the usual formal treatment[3] of doubly-periodic functions, we shall obtain explicit formulas for our $z(w)$ and $W(w)$. To this end, we first use a suitable linear

---

[3] Cf., for instance, Bieberbach, *Lehrbuch der Funktionentheorie*, Vol. 1 (Chelsea, 1945), or Knopp, *Theory of Functions*, Vol. 2 (Dover, 1950).

transformation in z that reduces the Riemann surface of two sheets, as well as the integral of the first kind defined on it, to an especially convenient form, called the **Weierstrass Normal Form.** The integral looks as follows in this normal form:

$$w = \int_{\infty}^{z} \frac{dz}{\sqrt{4z^3 + g_2 z + g_3}}.$$

One branch-point of the corresponding Riemann surface is at ∞; the remaining three have $z = 0$ as their centroid (i.e. their sum is zero), since the coefficient of z^2 under the radical is zero. After this "normalization," we find that

$$z(w) = \wp(w) = \frac{1}{w^2} + \sum_{h,k} \left[\frac{1}{w - \omega^2} - \frac{1}{\omega^2} \right],$$

$$\sqrt{4z^3 + g_2 z + g_3} = \wp'(w) = -\frac{2}{w^3} - 2 \sum_{h,k} \frac{1}{(w-\omega)^3},$$

$$\omega = 2h\omega_1 + 2k\omega_2, \; h^2 + k^2 \neq 0.$$

These are exactly the standard doubly-periodic functions, as studied in Bieberbach Vol. 1 or Knopp Vol. 2 (cf. footnote 3 of this section). To see this, we recall that first of all, $z(w)$ is a doubly-periodic function with a pole at $w = 0$; for, the branch-point a_4 that was mapped onto $w = 0$ is now at $z = \infty$. Therefore if we succeed in determining the principal part[4] of the power

---

[4] The *principal part* of a Laurent expansion is the sum of all the terms with negative exponents.

series expansion (Laurent expansion) of $z(w)$ in the neighborhood of this pole, then we shall have determined $z(w)$ to within an additive constant, according to the developments in Bieberbach Vol. 1 or Knopp Vol. 2. To obtain this principal part, we start from the expansion of $w(z)$ at $z = \infty$, viz.

$$w(z) = -z^{-\frac{1}{2}}\left(1 - \frac{g_2}{40}z^{-2} + \cdots\right),$$

and from this we obtain

$$\frac{1}{w^2} = z\left(1 + \frac{g_2}{20}z^{-2} + \cdots\right).$$

Therefore the desired principal part is $1/w^2$, since $\lim\limits_{z\to\infty}\left(\frac{1}{w^2} - z\right) = 0$. Hence $z(w) - \wp(w)$ is a doubly-periodic function without poles, and must therefore be a constant c. To prove that this constant is zero, we start by considering once more the limit $\lim\limits_{w\to 0}\left(\frac{1}{w^2} - z\right)$. The value of this limit was found above to be zero.

We may further conclude from the above expression for $\wp(w)$, that $\lim\limits_{w\to 0}\left(\wp(w) - \frac{1}{w^2}\right) = 0$. But since we set $z(w) - \wp(w) = c$, we now find that

$$c = \lim_{w \to 0} (z(w) - \wp(w)) = \lim_{w \to 0} \left(z(w) - \frac{1}{w^2} - \left(\wp(w) - \frac{1}{w^2} \right) \right)$$

$$= \lim_{w \to 0} \left(z(w) - \frac{1}{w^2} \right) - \lim_{w \to 0} \left(\wp(w) - \frac{1}{w^2} \right) = 0\,.$$

We thus have the following result:

$$z(w) = \wp(w) = \frac{1}{w^2} + \sum \left[\frac{1}{(w-\omega)^2} - \frac{1}{\omega^2} \right].$$

This is therefore the explicit expression for the function that *maps the rectangle with vertices* $0, \omega_1, \omega_1 + \omega_2, \omega_2$ *onto an upper half-plane bounded by the real axis of the z-plane*; an application of this will be made later.

Let us also express $\sqrt{4z^3 + g_2 z + g_3}$ as a func- of w. From

$$\int_\infty^z \frac{dz}{\sqrt{4z^3 + g_2 z + g_3}}$$

we find, by differentiation, that

$$\frac{dw}{dz} = \frac{dz}{\sqrt{4z^3 + g_2 z + g_3}}$$

and hence that

$$\sqrt{4z^3 + g_2 z + g_3} = \frac{dz}{dw} = \wp'(w) = \frac{-2}{w^3} - 2 \sum \frac{1}{(w-\omega)^3}\,.$$

The details of the mapping of the Riemann surface by an integral of the first kind for the case that the branch-points do not all lie on a circle will not be discussed here, beyond the mention of the following facts: If a suitable system of cuts is chosen for the Riemann surface, then its image under the mapping is a period-parallelogram; in this case, too, $z(w)$ and the radical are doubly-periodic functions; and in this case, too, the integral of the first kind gives the solution of the corresponding problem of uniformization.

CHAPTER FIVE

Mappings of Given Regions

§ 15. The Mapping of a Given Region Onto the Interior of a Circle (Illustrative Examples)

In his dissertation of 1851, Riemann stated a theorem to the effect that every simply-connected region having at least two boundary points can be mapped one-to-one and isogonally onto the interior of a circle. The proof of this theorem will occupy us in the subsequent sections; in the present section we are merely going to assemble various examples of such mappings, as applications of the developments of the preceding chapter. In some cases we shall obtain formulas for mapping the given region onto the interior of the unit circle, in other cases for mapping it onto the upper half-plane bounded by the real axis. There is no essential difference between these two, since we may always call on the function $w = (1 + iz)/(1 - iz)$ to map the upper half-plane $\mathfrak{I}z > 0$ onto the interior of the circle $|w| < 1$.

1. $w = z^n$ maps the **sector** $0 < \varphi < \dfrac{2\pi}{n}$, $z = re^{i\varphi}$

onto the upper half-plane $\mathfrak{I}w > 0$.

2. The **semi-circle** $|z| < 1$, $y > 0$ $(z = x + iy)$ is mapped onto the upper half-plane by the function

$w = \left(\frac{z+1}{z-1}\right)^2$; for, $\mathfrak{z} = (1+z)/(1-z)$ maps the given region onto the sector $0 < \varphi < \frac{\pi}{2}$, $\mathfrak{z} = re^{i\varphi}$, and this in turn is mapped as in 1. above.

3. The **circular sector** $|z| < 1$, $0 < \varphi < \pi/n$, is mapped onto the upper half-plane by

$$w = \left(\frac{z^n+1}{z^n-1}\right)^2$$

4. A **convex lens** formed by two circular arcs that intersect at an angle π/n at the points $z = a$ and $z = b$, is mapped onto a half-plane by

$$w = \left(\frac{z-a}{z-b}\right)^n$$

5. The **infinite strip** $0 < y < \pi$ $(z = x + iy)$ is mapped onto the upper half-plane by $w = e^z$, as follows from the developments of § 13.

6. We can now deal with the mapping of a **closed crescent** formed by two circles tangent to each other, say at $z = 0$, one of the circles being contained in the other. The function $w = 1/z$ maps the crescent onto an infinite strip bounded by two parallel lines. This strip can be mapped onto the one in 5. above by a rigid motion followed by a magnification, after which we proceed as in 5.

7. We can also map an **infinite half-strip** onto a half-plane. Such a half-strip is obtained by adding to the boundary of a doubly infinite strip, such as the one in 5. above, a line-segment joining, and perpendicular to, the two parallels that bound the strip, and then taking one of the two halves of the resulting figure. Since the exponential function maps the perpendiculars to the direction of the strip onto circles with center at $w = 0$, it maps the half-strip onto a semi-circular region, and then we can let 2. above take over.

8. The **rectangle** whose vertices are at $0, \omega_1, \omega_2, \omega_1 + \omega_2$ (ω_1 real and positive, ω_2 a positive pure imaginary) is mapped by

$$w = \wp(z; \omega_1, \omega_2) = \frac{1}{z^2} + \sum_{h,k} \left[\frac{1}{(z-\omega)^2} - \frac{1}{\omega^2} \right]$$

$$\begin{pmatrix} \omega = 2\,h\omega_1 + 2\,k\omega_2 \\ h, k = 0, 1, 2, \ldots\, h^2 + k^2 \neq 0 \end{pmatrix}$$

onto the upper half-plane (cf. § 14 above).

9. The **parabola** $y^2 = 4c^2(x + c^2)$ has $z = x + iy = 0$ as its focus. Its **exterior** region, the one not containing $z = 0$, is mapped by $w = \sqrt{z}$ onto the upper half-plane bounded by the straight line $\eta_2 = c > 0\,(w = \eta_1 + i\eta_2)$ (cf. § 11).

10. The same function maps the upper half (i.e. the half above the real axis) of the **interior** of the same **parabola** onto a half-strip, bounded by halves of the lines $\eta_2 = c > 0$ (corresponding

to the parabolic arc) and $\eta_2 = 0$ (corresponding to the positive real axis) and by a segment of the line $\eta_1 = 0$ (corresponding to the segment from vertex to focus) (cf. Fig. 33).

The half-strip can then be mapped onto the semi-circular region of Fig. 34 by means of $\mathfrak{z} = -e^{-\frac{\pi w}{c}}$. The radius of the circle equals 1; the vertex of the parabola corresponds to $\mathfrak{z} = +1$;

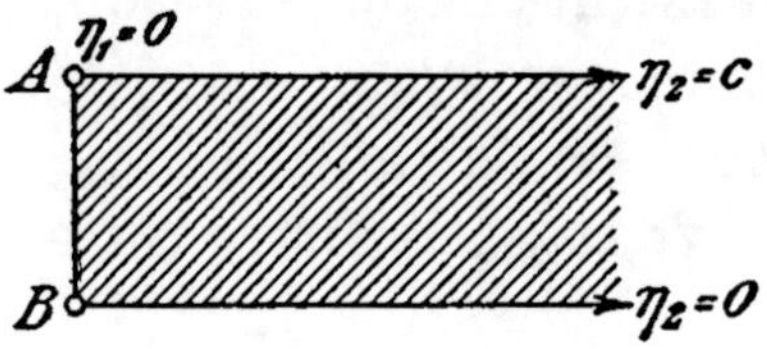

FIG. 33

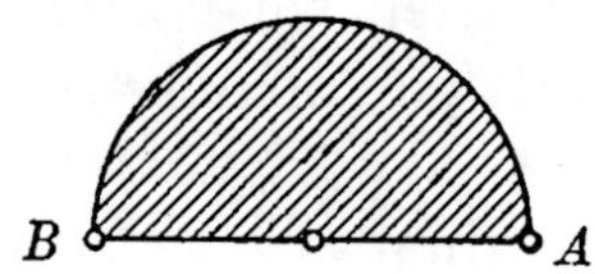

FIG. 34

$w = \infty$ gives $\mathfrak{z} = 0$.

Now consider the function[1]

$$\zeta = \left(\frac{\mathfrak{z}+1}{\mathfrak{z}-1}\right)^2 = \tanh^2\left(\frac{\pi\sqrt{z}}{2c}\right),$$

which maps the upper half of the interior of the parabola onto an upper half-plane (cf. Fig. 35).

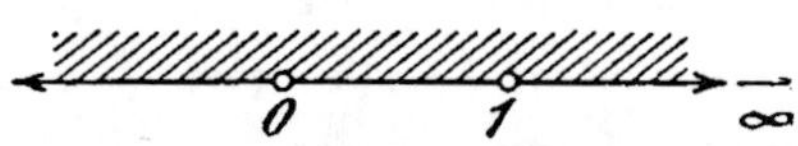

FIG. 35

[1] Recall that $\sinh z = (e^z - e^{-z})/2$, $\cosh z = (e^z + e^{-z})/2$, $\tanh z = \sinh z / \cosh z$.

The half-line from 1 to ∞ is the image of the parabolic arc itself, while the half-line from 1 *via* 0 to ∞ corresponds to the half-line that completes the boundary of the half of the interior. Hence by Schwarz' Reflection Principle, the entire interior of the parabola is mapped onto the whole plane, cut from 1 to ∞, by the function $\zeta = \tanh^2\left(\frac{\pi\sqrt{z}}{2c}\right)$. This plane is mapped in turn onto the upper half-plane by the function

$$w = \frac{1}{\sqrt{\zeta - 1}} = i\cosh\left(\frac{\pi\sqrt{z}}{2c}\right).$$

We thus have the following result: The interior of the parabola $y^2 = 4c^2(x + c^2)$ (where $z = x + iy$) is mapped onto the upper half-plane bounded by the real axis of the w-plane, by means of the function $w = i\cosh\left(\frac{\pi\sqrt{z}}{2c}\right)$.

11. The regions bounded by **ellipses** or **hyperbolas** can be mapped onto a half-plane by methods similar to the above. We have to call on the function $w = z + 1/z$, studied in § 12, whose inverse maps the **region bounded by the two branches of the hyperbola**

$$(1) \quad \frac{u^2}{(2\cos a)^2} - \frac{v^2}{(2\sin a)^2} = 1 \left(w = u + iv,\ 0 < a < \frac{\pi}{2}\right)$$

onto the sector $a < \varphi < \pi - a$ or onto the sector $a + \pi < \varphi < 2\pi - a$, depending on which branch of the inverse function $z = \frac{w + \sqrt{w^2 - 4}}{2}$ is used for the mapping. To obtain the first-mentioned sector as the image, we must choose the sign of the radical in such a way that $w = 0$ goes into $z = i$. This sector can then be mapped onto the upper half-plane of the $\mathfrak{z}$-plane. The final result for the mapping function is

$$\mathfrak{z} = \left(e^{-ia} \frac{w + \sqrt{w^2 - 4}}{2}\right)^{\frac{\pi}{\pi - 2a}}.$$

12. To map the **interior of a branch of the hyperbola** (1) onto the interior of the unit circle, we start with the half of the given region that lies above the real axis of the w-plane, as we also did in § 12. Let this half be the one located in $u > 0$, $v > 0$. The inverse of $w = z + 1/z$ maps it onto the sector $-a < \varphi < 0$, $|z| < 1$, provided that the branch of $z = \frac{w + \sqrt{w^2 - 4}}{2}$ that is used is the one for which $\sqrt{w^2 - 4} < 0$ if $w > 2$. By reflection in $|z| = 1$, we then see that the whole interior of the hyperbola, slit along $w > 2$, is mapped onto the sector $-a < \varphi < 0$, where $\varphi = -a$ corresponds to the hyperbola itself, and $\varphi = 0$ to the slit. Two points with the same w on opposite banks of the slit correspond to z and $1/z$

on $\varphi = 0$. We now map the sector by means of $\mathfrak{z} = z^{\frac{\pi}{2a}}$; then the new image of the interior of the hyperbola slit along $w > 2$ is the quadrant $-\pi/2 < \arg \mathfrak{z} < 0$. Again, any two points $\mathfrak{z}$ and $1/\mathfrak{z}$ lying in this quadrant correspond to the same w of the slit. Next, $\mathfrak{w} = (\mathfrak{z} - 1)/(\mathfrak{z} + 1)$ maps the quadrant onto the semi-circular region $|\mathfrak{w}| < 1$, $\Im \mathfrak{w} < 0$, in such a way that $-1 < \mathfrak{w} < 1$ corresponds to the slit and that any two points of this segment that differ only in sign correspond to the same w of the slit. Finally, $\mathfrak{w}^2$ completes the mapping of the interior of the branch of the hyperbola onto the interior of the unit circle.

13. The **exterior of the ellipse**

$$\frac{u^2}{a^2} + \frac{v^2}{b^2} = 1,\ a = r + \frac{1}{r},\quad b = \frac{1}{r} - r,\ r < 1$$

is mapped onto the interior of the circle $|z| = r < 1$ by the inverse of $w = z + 1/z$ (where $w = u + iv$).

14. The same function maps the half above the real axis of the **interior of the same ellipse** onto a quadrangular region, bounded by the semi-circles $|z| = 1$ and $|z| = r < 1$ above the real axis and by two segments of the real axis, provided that we use that branch of the inverse function

$$z = \frac{w + \sqrt{w^2 - 4}}{2}$$

which maps $w=0$ onto $z=i$. This quadrangular region is then mapped onto the rectangle $\log r < \Re \mathfrak{z} < 0$, $0 < \Im \mathfrak{z} < \pi$ by the function $\mathfrak{z} = \log z$, and the rectangle is mapped in turn onto a half-plane by means of $\wp(\mathfrak{z}, \log r, i\pi)$.

15. **Polygons.** The half-plane $\Im z > 0$ is mapped by

$$w = \int_0^z \frac{dt}{(t-a_1)^{\alpha_1}(t-a_2)^{\alpha_2}(t-a_3)^{\alpha_3}}, \quad \begin{array}{l} \alpha_1+\alpha_2+\alpha_3=2 \\ \alpha_1>0, \alpha_2>0, \alpha_3>0 \end{array}$$

onto (the interior of) a **triangle** with the exterior angles $\alpha_1\pi, \alpha_2\pi, \alpha_3\pi$. We can verify this by a discussion similar to the one on p. 96 ff. concerning the elliptic integral of the first kind, by observing that on each segment joining two of the three points a_1, a_2, a_3, the amplitude of the integrand is constant. The above formula is a special case of the general *Schwarz-Christoffel formula*

$$w = \int_0^z \frac{dt}{\Pi(t-a_k)^{\alpha_k}}, \quad \Sigma\alpha_k = 2,$$

which represents the mapping of $\Im z > 0$ onto (the interior of) a **polygon** with the exterior angles $a_k\pi$.

Using in addition a linear transformation on t, it is easy to see that integrals of a similar structure also give the mapping of $|z| < 1$ onto a polygon in the w-plane. As a special case, we mention the mapping of $|z| < 1$ onto a **regular n-gon** by means of

$$w = \int_0^z \frac{dt}{(1-t^n)^{\frac{2}{n}}}.$$

The α_k in the Schwarz-Christoffel formula are determined by the angles of the polygon, while the a_k depend on the lengths of the sides of the polygon. We must also allow for a constant factor before the integral sign, and for an additive constant, if we want the formula to include all possible polygons. By a suitable linear transformation on t, we can always make three of the a_k take on preassigned values. For example,

$$w = a \int_0^z \frac{dt}{(t+1)^{\alpha_1} t^{\alpha_2} (t-1)^{\alpha_3}} + b$$

gives the *general* mapping of $\Im z > 0$ onto a triangle with the exterior angles $\alpha_k \pi$. To obtain a given *specific* such triangle, we first determine $|a|$ in such a way that one side of the triangle will have its assigned length, and b and $\arg a$ are then determined in such a way that the triangle will assume its assigned position in the w-plane.

We note further that

$$w = \int_0^z \frac{(t-a_1)^{\alpha_1} \cdots (t-a_n)^{\alpha_n}}{t^2}\, dt, \quad \Sigma \alpha_k = 2, \quad \Sigma a_k \alpha_k = 0$$

maps the region $|z| > 1$ onto the *exterior* of a polygon with the exterior angles $\alpha_k \pi$. Here we have assumed $|a_k| = 1$. Similarly,

$$w = \int_0^z \frac{(t - a_1)^{\alpha_1} \cdots (t - a_n)^{\alpha_n}}{(1 + t^2)^2} dt$$

maps the half-plane $\Im z > 0$ onto the exterior of a polygon whose exterior angles are the $\alpha_k \pi$.

The elliptic integral of the second kind,

$$w = \int_0^z \sqrt{\frac{t}{(t - e_1)(t - e_2)}} dt,$$

maps the half-plane $\Im z > 0$ onto the domain indicated in Fig. 36. The corners are the images

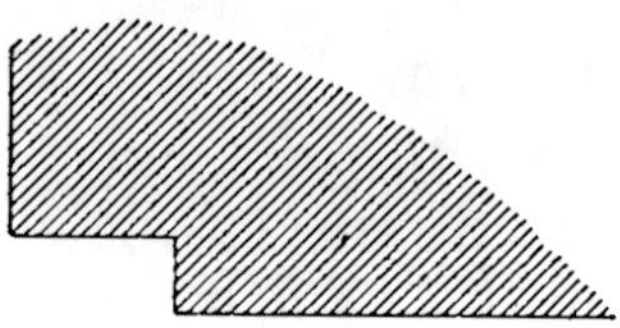

FIG. 36

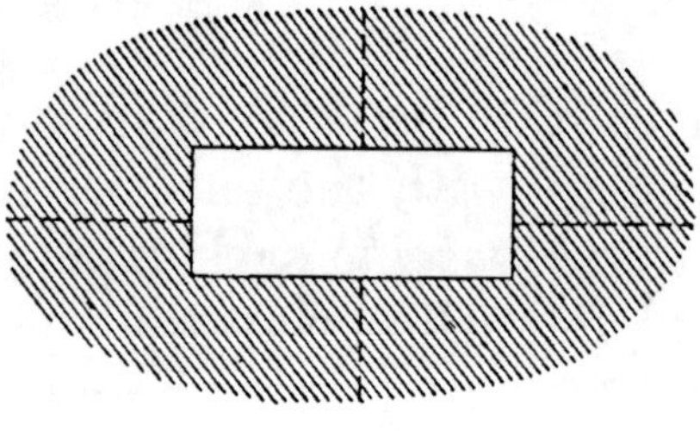

FIG. 37

of the points ∞, -1, 0, 1 of the real axis of the z-plane. It can be seen by repeated applications of reflection that the Riemann surface of two sheets with branch-points at ∞, -1, 0, 1

is mapped by the integral of the second kind onto the exterior of a rectangle (cf. Fig. 37). By combining this mapping with the one we studied in § 14, given by the elliptic integral of the first kind, we obtain the mapping of a half-plane onto the exterior of a rectangle.

16. Slit domains. By § 12, the function $\mathfrak{w} = z + 1/z$ maps the unit circle $|z| < 1$ onto the $\mathfrak{w}$-plane slit along $-2 < \Re\mathfrak{w} < +2$. Combining this mapping with $w = 1/(\mathfrak{w} + 2)$, we see that $w = z/(1+z)^2$ maps $|z| < 1$ onto the w-plane slit along $\Re w > 1/4$.

The function

$$w = \frac{1 + i\mathfrak{w}}{1 - i\mathfrak{w}}, \quad \mathfrak{w} = z + \frac{1}{z}$$

maps the unit circle $|z| < 1$ onto a region that is obtained from the full w-plane by omitting the points of a slit along part of the circle $|w| = 1$. This slit goes from $(-3 + 4i)/5$ *via* 1 to $(-3 - 4i)/5$.

§ 16. Vitali's Theorem on Double Series

In § 18 we shall give a proof of the Riemann Mapping Theorem, a theorem that has already been stated, and illustrated by examples, in § 15. The proof will be especially simple if we first take up a general convergence theorem.

The reader is familiar with Weierstrass' Theorem on Double Series,[1] which we state here as follows: Let the functions $f_\nu(z)$ be regular in $|z| < r$, and let $\Sigma f_\nu(z)$ be uniformly convergent in $|z| \leqq \varrho < r$, for every $\varrho < r$. Then $f(z) = \Sigma f_\nu(z)$ is regular in $|z| < r$.

Several authors—especially Stieltjes, Osgood, Vitali, and Montel—have modified the hypotheses of this theorem, still obtaining uniform convergence as the conclusion. The most far-reaching result in this direction is **Vitali's Double-Series Theorem:** *Let* $s_n(z) = \sum_0^n f_\nu(z)$. *For all* n, *let* $|s_n(z)| \leqq M$ *in* $|z| < r$, *where* M *is independent of* n. *Let* $\Sigma f_\nu(z) = \lim_{n\to\infty} s_n(z)$ *converge at the points* $z = z_\varkappa (\varkappa = 1, 2, \ldots)$ *that have a point of accumulation at* $z = 0$. *Then* $\Sigma f_\nu(z)$ *converges uniformly in* $|z| \leqq \varrho < r$, *for every* $\varrho < r$.

Proof. Take ϱ arbitrary in $0 < \varrho < r$, and then choose ϱ_1 with $\varrho < \varrho_1 < r$. Set $\lim_{n\to\infty} s_n(z_\varkappa) = s(z_\varkappa)$. We have for all ν that

$$|s_\nu(z) - s_\nu(0)| \leqq |s_\nu(z)| + |s_\nu(0)| < 2M.$$

Hence by Schwarz' Lemma (cf. p. 40), we have

---

[1] Cf. Carathéodory, *The Theory of Functions*, Vol. 1 (Chelsea, 1953), or Bieberbach, *Lehrbuch der Funktionentheorie*, Vol 1 (Chelsea, 1945), or Knopp, *Theory of Functions*, Vol. 1 (Dover, 1950).

$$|s_\nu(z_\varkappa) - s_\nu(0)| \leqq \frac{2M}{r}|z_\varkappa|.$$

Now assign $\varepsilon > 0$ arbitrarily, pick a $|z_\varkappa| < \frac{\varepsilon r}{4M}$, and then choose N such that for $\nu \geqq N$,

$$|s_\nu(z_\varkappa) - s(z_\varkappa)| < \frac{\varepsilon}{2}$$

holds.
Then we have

$$\begin{aligned}|s_\nu(0) - s(z_\varkappa)| &= |s_\nu(0) - s_\nu(z_\varkappa) + s_\nu(z_\varkappa) - s(z_\varkappa)| \\ &\leqq |s_\nu(0) - s_\nu(z_\varkappa)| + |s_\nu(z_\varkappa) - s(z_\varkappa)| \\ &< \frac{\varepsilon}{2} + \frac{\varepsilon}{2} = \varepsilon.\end{aligned}$$

Hence if $\mu \geqq N$, $\nu \geqq N$, then

$$|s_\nu(0) - s_\mu(0)| < 2\varepsilon,$$

and hence

$$\lim_{\nu\to\infty} s_\nu(0) = s_0$$

exists, and we have $|s_0| \leqq M$. We now use mathematical induction; set

$$s_\nu(z) = \Sigma s_{\nu\lambda} \cdot z^\lambda$$

and assume that for $\lambda = 0, 1, \ldots, n$, the existence of the limits

$$\lim_{\nu\to\infty} s_{\nu\lambda} = s_\lambda$$

and the inequality $|s_\lambda| \leqq \frac{M}{r^\lambda}$ have already been proved. We shall then prove them for $\lambda = n + 1$. To this end, consider the functions

$$S_\nu(z) = \frac{s_\nu(z) - \sum_0^n s_{\nu\lambda} z^\lambda}{z^{n+1}}.$$

These are regular in $|z| < r$. Since we have $|s_{\nu\lambda}| \leqq \frac{M}{r^\lambda}$ in $|z| < r$, by Cauchy's inequality on the coefficients, and since the maximum in $|z| \leqq \varrho_1$ of $S_\nu(z)$ is assumed on $|z| = \varrho_1$, it follows that

$$|S_\nu(z)| \leqq \frac{M + \sum_0^n \frac{M}{r^\lambda} r^\lambda}{\varrho_1^{n+1}}, \text{ in } |z| \leqq \varrho_1 .$$

This holds for all $\varrho_1 > \varrho$, so that we have for $\varrho_1 \to r$ that

$$|S_\nu(z)| \leqq \frac{(n+2)M}{r^{n+1}}, \text{ in } |z| < r .$$

We further have the existence of

$$\lim_{\nu \to \infty} S_\nu(z_\varkappa) \text{ for } \varkappa = 1, 2 \cdots .$$

Hence by an argument such as was just used above, we obtain

$$\lim_{\nu \to \infty} S_\nu(0) = \lim_{\nu \to \infty} s_{\nu, n+1} = s_{n+1} ,$$

and we have $|s_{n+1}| \leqq \frac{M}{r^{n+1}}$, since we have

$$|s_{\nu,n+1}| \leqq \frac{M}{r^{n+1}}$$

for all ν by Cauchy's inequality on coefficients.

Therefore the series

$$\sum_0^\infty s_\nu z^\nu$$

represents a function $s(z)$ regular in $|z| < r$. Now consider

$$s(z) - s_\nu(z) = \sum_0^\lambda (s_\mu - s_{\nu\mu}) z^\mu + \sum_{\lambda+1}^\infty (s_\mu - s_{\nu\mu}) z^\mu$$

in $|z| \leqq \varrho$. Since $|s_\mu| \leqq \frac{M}{r^\mu}$ and $|s_{\nu\mu}| \leqq \frac{M}{r^\mu}$, we have

$$\left| \sum_{\lambda+1}^\infty (s_\mu - s_{\nu\mu}) z^\mu \right| \leqq 2M \sum_{\lambda+1}^\infty \left(\frac{\varrho}{r}\right)^\mu \quad \text{for } |z| \leqq \varrho.$$

We can therefore choose λ in such a way that

$$2M \sum_{\lambda+1}^\infty \left(\frac{\varrho}{r}\right)^\mu < \frac{\varepsilon}{2}$$

holds, for any pre-assigned $\varepsilon > 0$. Having chosen such a λ, we then take ν so large that

$$\left| \sum_0^\lambda (s_\mu - s_{\nu\mu}) z^\mu \right| < \frac{\varepsilon}{2}$$

holds. This is possible because $\lim\limits_{\nu\to\infty} s_{\nu\mu} = s_\mu$ for all μ. Hence for sufficiently large ν, we have

$$| s(z) - s_\nu(z) | < \varepsilon$$

in $| z | \leqq \varrho$, and Vitali's Double-Series Theorem is proved.

The method of analytic continuation by chains of circles[2] leads immediately to the following generalization:

Let the $f_\nu(z)$ be single-valued and regular in a region R; set $\sum\limits_0^\mu f_\nu(z) = s_\mu(z)$ and let $| s_\mu(z) | < M$ in R, for all μ. Further, let $\Sigma f_\nu(z_\varkappa)$ be convergent at points $z_\varkappa$ having a point of accumulation in the interior of the region. Then $\Sigma f_\nu(z)$ converges uniformly in every closed subset of the interior of R.

COROLLARY. *Let there be given a sequence of functions $s_n(z)$, regular in a region R and such that $| s_n(z) | \leqq M$ for all n, with M independent of n. Then it is possible to select a subsequence of the $s_n(z)$ that converges uniformly to a limit function in every closed sub-region of R.*

Proof. Choose z_0 arbitrary in R, and select any infinite sequence of points $z_\varkappa$ for which $\lim\limits_{\varkappa \to \infty} z_\varkappa = z_0$. By a well-known theorem (cf. e.g. Knopp, Vol. 1), the numbers $s_n(z_1)$ have at least one point of accumulation; let

$$s_{1,1}(z), \; s_{2,1}(z) \cdots$$

[2] Cf. the texts referred to in footnote 1 of this section.

be a subsequence of the $s_n(z)$ for which $\lim_{n\to\infty} s_{n,1}(z_1)$ is one of these points of accumulation. From this subsequence we then select another subsequence that converges at z_2 as well, say

$$s_{1,2}(z),\ s_{2,2}(z)\cdots$$

Continuing this process, we can construct the "diagonal sequence" $s_{1,1}(z),\ s_{2,2}(z)\cdots$, which converges at every one of the $z_\varkappa$; for from its n-th term on, it is a subsequence of the sequence $s_{1,n}(z)\cdots$, which converges at each of the points $z=z_1,\ z=z_2,\ \ldots,\ z=z_n$, and this holds for any n. Hence the diagonal sequence satisfies the hypotheses of Vitali's Double-Series Theorem, and the conclusion concerning uniform convergence therefore applies, Q.E.D.

§ 17. A Limit Theorem for Simple Mappings[1]

Let there be given a region R of the z-plane, and a sequence of functions $f_1(z), f_2(z), \ldots$ each of which is single-valued and regular in R and each of which gives a simple mapping[1] *of R. Assume further that on every closed sub-region*

[1] We have defined earlier what is meant by a *simple region* (cf. p. 4). A *mapping* is called *simple* (or schlicht) if the image region is simple. This means that the *mapping function* assumes each of its values only once, i.e. that *its inverse is single-valued.* [*Trans.*]

of R, the $f_n(z)$ converge uniformly to a limit function $f(z)$ that is not a constant in R. Then $f(z)$ also maps R simply.

Proof. Assume the conclusion to be false. Then there would be two different points z_1 and z_2 in R at which $f(z_1) = f(z_2) = a$, say. About each of the two points as centers, we draw a circle of radius δ, where δ is chosen so small that the two circles do not intersect and that neither circle contains within or on it any further point, besides its center, at which $f(z) = a$. Let $\varepsilon > 0$ be a number such that $|f(z) - a| > \varepsilon$ holds on the periphery of each of the circles. Then choose n large enough for $|f_n(z) - f(z)| < \varepsilon$ to hold on the periphery of each circle.

We can now prove that for n large enough, $f_n(z)$ would have to take on the value a at two different points of R, one in each of the two above circles, which will prove our theorem, considering that we shall then have a contradiction to the simpleness, assumed in the theorem, of the mapping $f_n(z)$ of R. The proof follows easily from Rouché's Theorem.[2] For the convenience of the reader, we shall here recapitulate the proof of Rouché's Theorem, in a form that applies directly to our present case. If the integral

[2] Cf. Carathéodory, *The Theory of Functions*, Vol. 1 (Chelsea, 1953), or Bieberbach, *Lehrbuch der Funktionentheorie*, Vol. 1 (Chelsea, 1945).

$$\frac{1}{2\pi i}\int \frac{f_n'(z)}{f_n(z)-\alpha}\,dz$$

is taken over the periphery of one of our circles in the positive sense, its value equals the number of α-places of $f_n(z)$ (i.e. of places z where $f_n(z)=\alpha$) within the circle. The integral is meaningful, since on the periphery of the circle the relation

$$\begin{aligned}|f_n(z)-\alpha| &= |f_n(z)-f(z)+f(z)-\alpha)| \\ &\geqq |f(z)-\alpha|-|f_n(z)-f(z)| \\ &> \varepsilon-\varepsilon=0\end{aligned}$$

holds. But

$$\frac{1}{2\pi i}\int \frac{f'(z)+\lambda[f_n'(z)-f'(z)]}{f(z)+\lambda[f_n(z)-f(z)]-\alpha}\,dz\,,$$

taken over the same circle, depends continuously on λ in $0\leqq\lambda\leqq 1$, since

$$\begin{aligned}|f(z)+\lambda[f_n(z)-f(z)]-\alpha| &\geqq |f(z)-\alpha|-\lambda|f_n(z-f(z)| \\ &> \varepsilon-\varepsilon=0.\end{aligned}$$

For $\lambda=1$, however, the last integral gives the number of α-places of $f_n(z)$ in the circle, and $\lambda=0$ those of $f(z)$. These two whole numbers must equal each other, since a continuous function of λ that only takes on whole numbers as its values, must be a constant. Hence if each of the two circles contained an α-place of $f(z)$, then each of them would also have to contain an α-place of

$f_n(z)$, for all sufficiently large n. Since the two circles do not intersect, this would contradict the simpleness of the mappings represented by these $f_n(z)$, Q.E.D.

§ 18. Proof of Riemann's Mapping Theorem

The theorem we wish to prove is as follows:

Every simple and simply-connected region R having at least two boundary points can be mapped one-to-one onto a circular disc by means of an analytic function.

We note first that we may confine ourselves to regions wholly contained in the interior of a finite circle. For if R is a region that is not finite, let a and b be two distinct boundary points of R. We map the Riemann surface of two sheets whose branch-points are at a and b onto the entire simple w-plane by means of the function $w = \sqrt{\dfrac{z-a}{z-b}}$. Since the simple region R may be thought of as forming a proper sub-region of the above Riemann surface, the image of R under the above mapping must fail to cover some region of the w-plane, and can therefore easily be mapped onto part of a finite circular disc, by means of a suitable linear function.

We therefore assume that we are given a finite region R, wholly contained within some circle.

Consider the set $\mathfrak{M}$ of all functions $f(z)$ regular in R whose modulus has a finite upper bound $\mu(f)$, depending on f, and which, furthermore, are such that every $f(z)$ gives a simple mapping of R, and such that, finally, at some fixed point z_0 of R, every $f(z)$ satisfies $f(z_0) = 0$, $f'(z_0) = 1$. Let ϱ be the greatest lower bound of all $\mu(f)$. Even if there were no function φ in $\mathfrak{M}$ for which $\mu(\varphi) = \varrho$, there must at any rate be a subset $f_1, f_2, \ldots$ of $\mathfrak{M}$ which is such that $\lim\limits_{n\to\infty} \mu(f_n) = \varrho$. In R, the moduli of these f_n must stay less than a fixed bound independent of n, since $\mu(f_n)$ can not exceed ϱ by very much for all large enough n.

Hence, by § 16, a suitable subsequence of the f_n converges uniformly to a limit function. To simplify the notation in what follows, let us take the f_n to be that convergent subsequence to begin with, and denote the limit function by $f(z)$. Since $f(z_0) = 0$, $f'(z_0) = 1$, the limit function is not a constant in R, and by § 17 it therefore represents a simple mapping of R. Furthermore, $\mu(f) \leqq \varrho$ holds, since $f(z)$ is the limit of functions f_n whose $\mu(f_n)$ exceeds ϱ by arbitrarily little if n is sufficiently large; therefore $f(z)$ belongs to the set $\mathfrak{M}$. Hence we have $\mu(f) = \varrho$, recalling that ϱ was defined as the greatest lower bound of the μ's for functions in $\mathfrak{M}$. We also have $\varrho > 0$.

We shall now prove that $w = f(z)$ maps the

region R onto the *full* circular disc $|w| < \varrho$; so far, we know merely that the image region does not contain any points outside this circle. Now if the image region were not identical with the whole interior of the disc, then this interior would have to contain boundary points of the image region; and this would imply, as we shall demonstrate presently, the existence of a function $\varphi(z)$ regular in R that gives a simple mapping of R and for which $\varphi(z_0) = 0$, $\varphi'(z_0) = 1$, and $\mu(\varphi) < \varrho$, contrary to the definition of ϱ. It obviously suffices to construct a corresponding function $\varphi(w)$ in the image region R' of R under the mapping $w = f(z)$. Let α, with $|\alpha| < \varrho$, be a boundary point of R', and recall that $0 = f(z_0)$. Then each branch of

$$w_1 = \varrho\sqrt{\frac{\varrho(w-\alpha)}{\varrho^2 - \bar{\alpha}w}}$$

is regular and simple in $|w| < \varrho$, and we have $\mu(w_1) = \varrho$, since any boundary point of R' lying on $|w| = \varrho$ is mapped onto a w_1 with $|w_1| = \varrho$. We decide on one of the branches of $w_1(w)$, and with this branch form

$$w_2 = \frac{\varrho^2(w_1 - w_1(0))}{\varrho^2 - \overline{w_1(0)}\,w_1}.$$

Now w_2, considered as a function of z, is regular and simple in R and satisfies $w_2(0) = 0$, as well as $\mu(w_2) = \varrho$ in R. But this function does not belong to $\mathfrak{M}$, since its derivative at $w = 0$ (i.e. at $z = z_0$) is $\neq 1$, being instead equal to

$$\frac{\varrho + |\alpha|}{2\sqrt{-\bar{\alpha}\varrho}}.$$

The modulus of this number exceeds unity. However, the function

$$w_3 = \frac{2w_2\sqrt{-\bar{\alpha}\varrho}}{\varrho + |\alpha|} = h(w, \varrho, \alpha)$$

belongs to the set $\mathfrak{M}$, and $\mu(w_3) < \varrho$ holds. This contradiction to the definition of ϱ shows that $w = f(z)$ must map R onto the full disc $|w| < \varrho$, and Riemann's Mapping Theorem is proved.

Remark. For simply-connected regions having only one boundary point, the above mapping theorem does not, of course, hold, because of Liouville's Theorem.[1]

§ 19. On the Actual Construction of the Conformal Mapping of a Given Region Onto a Circular Disc

It may be considered a drawback of the proof in the preceding section that it appeared to give no method for *actually determining* the mapping

[1] Cf. Carathéodory, *The Theory of Functions*, Vol. 1, § 167 (Chelsea, 1953); Bieberbach, *Lehrbuch der Funktionentheorie*, Vol. 1 (Chelsea, 1945); Knopp, *Theory of Functions*, Vol. 1 (Dover, 1950).

whose existence it assures. We shall see, however, that a repeated use of the function $h(w, \varrho, \alpha)$ of the preceding section *does* lead to such a method. To this end, assume that the region R to be mapped contains the point $z = 0$ and is contained in the disc $|z| < \varrho$. Let $z = \alpha$ be (that) one of its boundary points which is closest to $z = 0$. Then $z_1 = h(z, \varrho, \alpha)$ maps R onto a region R_1 that contains $z_1 = 0$ and is contained in a circular disc $|z_1| < \varrho_1 < \varrho$. Let $z_1 = \alpha_1$ be one of the boundary points of R_1 closest to $z_1 = 0$; then we can prove that $\frac{|\alpha_1|}{\varrho_1} > \frac{|\alpha|}{\varrho}$. To see this, we need merely observe that under the mapping of R by

$$z' = h(z, \varrho, \alpha) \cdot \frac{\varrho + |\alpha|}{2\sqrt{-\bar{\alpha}\varrho}},$$

the boundary point α' closest to $z' = 0$ satisfies the relation $|\alpha'| > |\alpha|$, which follows from Schwarz' Lemma applied to the inverse function of $z'(z)$.

The above hint should suffice to indicate that repeated applications of the procedure will lead to mappings of R under which the image of the complete boundary of R lies in a circular annulus whose radii are in a ratio differing from unity by as little as we please.

This iterative method of mapping is called the *method of osculation*. It has not so far found practical use, although the individual steps can easily be carried out by graphical methods. The reason is that a more detailed study reveals the convergence of the process to be in general rather slow.

A method due to F. Ringleb that is useful in actual practice consists in replacing the mappings of circles that figure in the above steps by mappings of convex lenses, or of crescents, that contain R. For if an arc of $|z| = \varrho$ is free of boundary points of R, then we can cut off a crescent, free of points of R, from $|z| < \varrho$. There remains a convex lens contained in $|z| < \varrho$ and containing R. This lens we map *onto* $|z| < \varrho$, as we did in § 15, taking care also to map the origin onto itself. By Schwarz' Lemma, this mapping increases the distance from the origin of every point of R. Because of the fact that we are now mapping onto a disc a smaller sub-region of $|z| < \varrho$ than we did above in the method of osculation, the boundary points of R are pushed toward the periphery of $|z| < \varrho$ faster than in the osculation method. Furthermore, the mappings involved in the new method are isogonal at every boundary point of R, in contradistinction to those involved in the osculation method, where corners are introduced. Hence the successive images of the boundary of R will in the new method give a superior approximation of the *direction* of $|z| = \varrho$ as well, so that the circumferences of the successive images will converge to that of the circle.

Another useful method is based on the following theorem, which is concerned with the behavior of area under conformal mappings:

THEOREM. *If*

$$f(z) = z + a_2 z^2 + \dots$$

maps the disc $|z| < r$ *onto a region* R, *then the area* J *of* R *is given by*

$$J = \pi(r^2 + 2\,|a_2|^2 r^4 + \cdots + n\,|a_n|^2 r^{2n} + \cdots),$$

so that J *always exceeds the area of the circle that is being mapped,* except in the trivial case $f(z) = z$.

Proof. We have

$$J = \int_0^r \int_0^{2\pi} |f'(z)|^2 \varrho \, d\varrho \, d\varphi,$$

where $z = \varrho e^{i\varphi}$, since as is well known—setting $f(z) = u + iv$, $z = x + iy$, as usual—the area is given by

$$J = \iint_{|z|<r} \begin{vmatrix} u_x & u_y \\ v_x & v_y \end{vmatrix} dx\,dy.$$

This last integral is equal to the one above by virtue of the Cauchy-Riemann differential equations. Since $|c|^2 = c\bar{c}$ holds for any complex number c, we have further that

$$J = \int_0^r \int_0^{2\pi} (1 + 2a_2 z + \cdots)(1 + 2\bar{a}_2 \bar{z} + \cdots) \varrho \, d\varrho \, d\varphi$$

$$= \int_0^r \int_0^{2\pi} (1 + 2a_2 \varrho e^{i\varphi} + \cdots)(1 + 2\bar{a}_2 \varrho e^{-i\varphi} \cdots) \varrho \, d\varrho \, d\varphi.$$

To evaluate this integral, we first carry out the integration with respect to φ, observing that the terms $e^{hi\varphi}$ obtained by multiplying out yield zero upon integration from $\varphi = 0$ to $\varphi = 2\pi$ except when $h = 0$. This leads to the expression for J given in the theorem, Q.E.D.

The result just proved can be used for the actual determination of the conformal mapping of a given region onto a circular disc. For, the theorem implies that of all conformal mappings of a given region R that map a given point of R onto zero, and that are such that the derivative of the mapping function at the given point equals unity, *the mapping onto a simple circular disc gives the image with the smallest possible area.* Thus if R lies in the w-plane and if the given point mentioned above is $w = 0$, then we must determine $z = \varphi(w) = w + c_2 w^2 + \cdots$ in such a way that

$$\iint_B |\varphi'(w)|^2 \, du \, dv$$

(where $w = u + iv$) is as small as possible. This can be done without too much difficulty by computation, provided that we confine ourselves to some fixed finite number of terms of the above power series $\varphi(w)$. The problem is then that of determining among all integral rational functions of a given degree, the one which maps R onto a region whose area is to be as small as possible. In actual practice, this method has been found very useful when applied to regions R that are not *too* far from circular to begin with. (Cf. for instance the paper by Höhndorf, Ztschr. f. angew. Math. u. Mechanik, Vol. 6, p. 265 ff.)

In conclusion, we shall state here without proof a theorem that permits an estimate of how well a given mapping of a region approximates the mapping onto a circular disc. The theorem is as follows: *Let R be a simple and simply-connected region whose boundary curve is of length L. Let $|L - 2\pi| < \varepsilon \cdot 2\pi$, where $0 < \varepsilon < 1$, and suppose that the boundary curve lies entirely in the annulus formed by the two concentric circles whose center is at $z = 0$ and whose radii are $1 - \varepsilon$ and $1 + \varepsilon$. Let $f(z)$ be the analytic function that gives a simple mapping of $|z| < 1$ onto R and that satisfies $f(0) = 0$, $f'(0) > 0$. Then the following relation holds in $|z| \leqq 1$ (i.e. within and on the unit circle):*

$$|f(z) - z| < 2.2 \cdot \pi \sqrt{2\varepsilon}.$$

This bound is exact insofar as $\sqrt{\varepsilon}$ can not be replaced by a higher power of ε.

For $|z| \leqq r < 1$, Schwarz' Lemma even implies that $|f(z) - z| < 2.2 \cdot \pi \cdot \sqrt{2\varepsilon} \cdot r$.

For the proof, see Bieberbach, Sitzungsber. Preuss. Akad. d. Wiss., 1924, p. 181, and Landau, Jahresber. d. deutsch. Math. ver. Vol. 34 (1926), p. 241 ff.

§ 20. Potential-Theoretic Considerations

A function $u(x, y)$ is called a *potential function*, or *harmonic function*, if it satisfies the differential equation

$$\Delta u \equiv \frac{\partial^2 u}{\partial x^2} + \frac{\partial^2 u}{\partial y^2} = 0.$$

As the Cauchy-Riemann differential equations (1) of p. 1 show, the real part of any analytic function $f(z) = u + iv$ $(z = x + iy)$ is a harmonic function. Conversely, as is well known from Function Theory, every harmonic function is the real part of an analytic function.[1]

We shall first take up a transformation of the Cauchy Integral Formula. Let there be given a simply-connected region D of the ζ-plane that has at least two boundary points, and let $w = f(\zeta, z)$ be an analytic function that gives a simple mapping of D onto $|w| < 1$ under which $\zeta = z$ goes into $w = 0$. We assume furthermore that the boundary of D is an analytic curve, and that $f(\zeta, z)$ is

[1] Cf. Carathédory, *The Theory of Functions*, Vol. 1, § 154 (Chelsea, 1953), or Bieberbach, *Lehrbuch der Funktionentheorie*, Vol. 1 (Chelsea, 1945).

analytic also on this boundary[2]—as was the case in the examples of § 15. Here, an analytic curve is a curve $\zeta = \zeta(t)$, $\alpha \leqq t \leqq \beta$ such that $\zeta(t)$ can be developed in the neighborhood of every $t = t_0$ in powers of $t - t_0$.

At $\zeta = z$, we have an expansion of the form

$$f(\zeta, z) = a_1(\zeta - z) + \cdots \quad (a_1 \neq 0).$$

Hence

$$\frac{f'(\zeta, z)}{f(\zeta, z)} = \frac{1}{\zeta - z} + f_1(\zeta, z). \tag{1}$$

Here $f_1(\zeta, z)$ is a regular function of ζ in the region and on its boundary.

Let there be given also a function $f(\zeta)$ that is regular in D and continuous in D plus its boundary. Then

$$f(z) = \frac{1}{2\pi i}\int f(\zeta)\frac{1}{\zeta - z}\,d\zeta, \tag{2}$$

where the integral is taken over the boundary of D in the positive sense, i.e. such that the interior remains to the left as the boundary is traversed. Because of (1), we obtain from (2) that

$$f(z) = \frac{1}{2\pi i}\int f(\zeta)\frac{f'(\zeta, z)}{f(\zeta, z)}\,d\zeta - \frac{1}{2\pi i}\int f(\zeta) f_1(\zeta, z)\,d\zeta.$$

The second integral on the right is zero, by Cauchy's Integral Theorem. Hence we have

---

[2] We shall soon see that the analyticity of the mapping on the boundary is a consequence of the analytic character of the boundary curve.

$$f(z) = \frac{1}{2\pi i}\int f(\zeta)\frac{f'(\zeta, z)}{f(\zeta, z)}\,d\zeta\,. \tag{3}$$

If we set

$$\log f(\zeta, z) = -(g + ih)\,,$$

then the real function g is called the *Green's function* belonging to the region D and the point z. It is a harmonic function that is regular in D except at the point z, i.e. the real part of an analytic function that is regular in D except at z. The function g is single-valued and differs at $\zeta = z$ from a regular function by the additional term $-\log|\zeta - z|$. At the boundary of the region we have $g = 0$, so that g is continuous on the closed set consisting of D plus its boundary, except at the point $\zeta = z$. This follows from the fact that $g = -\log|f(\zeta, z)|$, that $|f(\zeta, z)|$ is continuous on the above closed set, and that $|f(\zeta, z)| = 1$ on the boundary.

Upon substituting

$$\log f(\zeta, z) = -(g + ih)$$

in (3), we obtain

$$\frac{f'}{f}\,d\zeta = -\left(\frac{\partial g}{\partial s} + i\,\frac{\partial h}{\partial s}\right)ds\,,$$

where $\dfrac{\partial}{\partial s}$ denotes differentiation with respect to arc-length. Here, the direction of increasing arc-length is the same as the positive direction of traversal, described in connection with (2) above.

But since we have seen that g is constant along the boundary, we must have $\frac{\partial g}{\partial s}=0$; and since we also know that $\frac{\partial g}{\partial n}=-\frac{\partial h}{\partial s}$, where $\frac{\partial g}{\partial n}$ denotes the derivative of g in the direction of the interior normal to the boundary,[3] we finally obtain

$$f(z)=\frac{1}{2\pi}\int f(\zeta)\frac{\partial g}{\partial n}ds.$$

If we separate out the real part u of $f(z)$, and set $\zeta=\xi+i\eta$, we find that

$$u(x,y)=\frac{1}{2\pi}\int u(\xi,\eta)\frac{\partial g}{\partial n}ds. \tag{4}$$

This is the so-called Green's Formula, which expresses the values taken on in D by a harmonic function regular in D and continuous in D plus its boundary, in terms of its values on the boundary (its "boundary values").

Let us apply this formula to the special case that D is the circular disc $|\zeta|<R$ and that $f(\zeta, z)$ maps this disc onto itself in such a way that $\zeta=z$ is mapped onto the center of the disc. Then we may set

$$f(\zeta,z)=\frac{R(\zeta-z)}{R^2-\zeta\bar{z}}$$

---

[3] The pair of axes ξ, η is oriented the same way as are s, n, so that ξ, η can be brought into coincidence with s, n by a proper rigid motion. It follows that $g+ih$ is an analytic function of $\zeta=s+in$.

Hence if $\zeta = Re^{i\vartheta}$, $z = re^{i\psi}$, then

$$\begin{aligned}\frac{f'}{f}\,d\zeta &= \frac{i\zeta}{\zeta - z}\,d\vartheta + \frac{\bar{z}i\zeta}{R^2 - \zeta\bar{z}}\,d\vartheta \\ &= \left(\frac{Re^{i\vartheta}}{Re^{i\vartheta} - re^{i\psi}} + \frac{re^{-i\psi}e^{i\vartheta}}{R - e^{i\vartheta}re^{-i\psi}}\right) id\vartheta \\ &= \left(\frac{Re^{i\vartheta}}{Re^{i\vartheta} - re^{i\psi}} + \frac{re^{-i\psi}}{Re^{-i\vartheta} - re^{-i\psi}}\right) id\vartheta \\ &= \frac{R^2 - r^2}{R^2 + r^2 - 2rR\cos(\vartheta - \psi)}.\end{aligned}$$

Therefore

$$(5)\qquad u(x, y) = \frac{1}{2\pi}\int_{\pi}^{+\pi} u(\vartheta)\frac{(R^2 - r^2)\,d\vartheta}{R^2 + r^2 - 2rR\cos(\vartheta - \psi)},$$

where $u(\vartheta)$ denotes the boundary values of u.

This special case of Green's Formula is called *Poisson's Integral Formula.*

If we take in particular $r = 0$, then (5) yields

$$(5a)\qquad u(0, 0) = \frac{1}{2\pi}\int_{-\pi}^{+\pi} u(\vartheta)\,d\vartheta,$$

which implies that *a harmonic function other than a constant can assume neither a maximum nor a minimum in the interior of its domain of regularity.* For if it had, say, a maximum at some interior point, then we could draw a circle with center at that point and with a suitable radius R such that the function assumes on the circumference values that never exceed, but some of which are less than, its value at the center, and we would have a contradiction to formula (5a).

Since the factor of $u(\vartheta)$ in the integrand of (5) is the real part of $\frac{\zeta+z}{\zeta-z}$, it follows that (5) always represents a harmonic function of x, y, provided that $|u(\vartheta)|$ is bounded and that

$$\int_{-\pi}^{+\pi} |u(\vartheta)|\, d\vartheta$$

exists, since under these assumption the integral

$$(6) \qquad \frac{1}{2\pi}\int_{-\pi}^{+\pi} u(\vartheta)\frac{\zeta+z}{\zeta-z}\,d\zeta$$

represents a regular analytic function of z for all $|z| < R$. This follows from Vitali's Double-Series Theorem if that theorem is applied to a sequence of approximating sums that converge to the integral (6).

If we can also show that under the above assumptions on $u(\vartheta)$ the Poisson Integral represents a function which *assumes* on the periphery the *given* boundary values $u(\vartheta)$, then we shall have demonstrated that the so-called *first boundary-value problem of potential theory* always has a solution in the case of a circle, i.e. that there always exist harmonic functions which are regular within the circle and which converge to the prescribed boundary value as a point ϑ of the boundary is approached. This will indeed be the case for every point ϑ where $u(\vartheta)$ is continuous, provided only that the approach to the point ϑ takes place within an angular sector whose vertex is at the point ϑ and all of whose remaining points in a neighborhood of the point ϑ lie in the interior of our circle. For every $\varepsilon > 0$, we can then cut from the angular sector a sector of a circle, with center at the point ϑ, so that within the circular sector,

$$|u(x, y) - u(\vartheta)| < \varepsilon$$

holds.

To prove the above statement, let us start by writing the Poisson Integral in the form (3), where we must substitute

$$f(\zeta, z) = \frac{R(\zeta - z)}{R^2 - \bar{z}\zeta} .$$

The values of the function f that enter here are those assumed on $|\zeta| = R$. But on this periphery, we have $|f| = 1$. Therefore we may set

$$f = e^{i\omega}, \ \omega \text{ real}.$$

Then

$$\frac{f'}{f}\frac{d\zeta}{ds} = \frac{d}{ds}\log f = i\frac{d\omega}{ds}$$

results, and the Poisson Integral can therefore be written also in the following form:

$$(7)\quad u(x, y) = \frac{1}{2\pi}\int_{-\pi}^{+\pi} U(\omega)d\omega,\ U(\omega) = u(\vartheta(\omega)).$$

To determine the significance of the angle ω, observe that $w = f(\zeta, z)$ maps the circle $|\zeta| < R$ onto $|w| < 1$ in such a way that $\zeta = z$ goes into $w = 0$; this shows that ω is the angle formed at z by a fixed initial direction and the circles passing through the two points z and R^2/z, the line elements (tangents) at z of the bundle of circles being taken up in the order corresponding to the positive sense of rotation about z. The circles in question may also be described as being those circles passing through the point z that are perpendicular to $|\zeta| = R$. We see from this that the mean-value formula (5a) is merely a special case of the more general transformed version (7). Now it is easy to derive from (7) the desired proof of our statement concerning the boundary values. To this end, consider a point ϑ_0 of continuity of $u(\vartheta)$, and an arc of the periphery containing this point and short enough for $|u(\vartheta) - u(\vartheta_0)| < \varepsilon$ to hold along this arc. If the point z is held fixed, then ω may be considered as a parameter along the periphery of the circle, varying from $-\pi$ to $+\pi$. Since $f(\zeta, z)$ may always be modified by any constant factor of modulus unity, we may assume that the point ϑ_0 corresponds to $\omega = 0$. Then if z is chosen sufficiently close to ϑ_0, and within a sufficiently small circular sector which is interior to the circle $|\zeta| = R$ except for its vertex at the point ϑ_0, then the above arc containing ϑ_0 will correspond to values of ω that include all of $-\pi + \varepsilon < \omega < \pi - \varepsilon$; also, $\varepsilon \to 0$ as $z \to \vartheta_0$. This means geometrically that if z is close to ϑ_0, then our mapping pulls the whole periphery close to ϑ_0, so to speak; this is certainly geometrically plausible.

Hence if we write

$$u(x, y) = \frac{1}{2\pi} \int_{-\pi+\varepsilon}^{\pi-\varepsilon} U(\omega) d\omega + \frac{1}{2\pi} \int_{-\pi}^{-\pi+\varepsilon} U(\omega) d\omega + \frac{1}{2\pi} \int_{\pi-\varepsilon}^{\pi} U(\omega) d\omega,$$

then, considering that $U(\omega)$ is bounded, the second of the integrals will go to zero with ε. The third will do likewise, while the first differs from

$$\frac{1}{2(\pi-\varepsilon)} \int_{-\pi+\varepsilon}^{\pi-\varepsilon} U(o) dw = U(o) = u(\vartheta_0)$$

by a number which goes to zero with ε. This completes our proof.

Thus *the Poisson Integral, in the case of bounded and absolutely integrable boundary values $u(\vartheta)$, represents a harmonic function that has, at every point ϑ_0 of continuity of $u(\vartheta)$, the* *boundary value* $u(\vartheta_0)$ *whenever the approach to that point is made within an angular sector all of whose other points are interior points of the given circle.*

Formula (7) shows once more that *in the case of bounded boundary values, the Poisson Integral represents a harmonic function whose values always are between the maximum and minimum of the boundary values.*

The arguments used in the above proof can also be applied to the neighborhood of a jump-discontinuity of the boundary-value function. If we approach such a point from either one of the two directions along the periphery, the boundary-value function will tend to a finite limit. Our arguments then show that *within a sufficiently small circular sector with vertex at the jump-discontinuity, the values taken on by the Poisson Integral will not exceed the larger of the two limits by more than* ε, *and will not be exceeded by the smaller limit by more than* ε.

If the boundary-value function is continuous along some arc of the periphery, then the function represented by the Poisson Integral is continuous on the point-set consisting of this arc plus the interior of the circle. This can easily be deduced from the above arguments, as follows. Let ϑ_0 be an interior point of the given arc of continuity of $u(\vartheta)$. Construct a quadrangular region bounded by parts of two radii symmetrically located with respect to ϑ_0, by the arc through ϑ_0 joining them, and by an arc of a slightly smaller concentric circle. We must show that if this quadrangular region is made sufficiently small, then the values of the harmonic function represented by the Poisson integral will differ from $u(\vartheta_0)$ by less than an arbitrarily pre-assigned number $\varepsilon > 0$. To show this, we start by choosing the angle between the two above radii so small that along the arc through ϑ_0 joining them along the periphery, $|u(\vartheta) - u(\vartheta_0)| < \frac{\varepsilon}{2}$ holds. We know from our arguments two pages back that every point ϑ of the arc is the vertex of some circular sector, interior to $|\zeta| = R$ except for its vertex, in which $u(x, y)$ differs from its boundary value $u(\vartheta)$ at the vertex by less than $\varepsilon/2$. In all these circular sectors, therefore, $u(x, y)$ differs from $u(\vartheta_0)$ by less than ε. As is apparent from our earlier proof just referred to, the radii of all these sectors can all be chosen so as to have a non-zero, positive lower bound which depends only on the maximum of the moduli of the boundary values. Hence by choosing the radial width of our quadrangular region to be less than this lower bound (i.e. by choosing the smaller concentric circle sufficiently close to $|\zeta| = R$), we can insure that every interior point of the quadrangular region also belongs to the interior of one of the above circular sectors, and that therefore $| u(x, y) - u(\vartheta_0) | < \varepsilon$ holds for every point (x, y) within the quadrangular region. We have thus proved also the continuity in the closed disc of the harmonic function represented by the Poisson Integral, in the case of a continuous boundary-value function $u(\vartheta)$.

It follows further that in the case of continuous boundary values, the harmonic function represented by the Poisson Integral is the *only* harmonic function that is continuous on the closed disc and assumes the given boundary values. For, the difference of two such functions would be a harmonic function continuous on the closed disc and having zero as its boundary values. If this function were not identically zero on the disc, it would assume its maximum and its minimum in the interior of the circle and would thus yield a contradiction to an earlier result of this section.

From this, it follows that a harmonic function that is continuous on a closed domain is always uniquely determined by its boundary values.

Under a conformal mapping of a region D of the ζ-plane onto a region D′ of the w-plane by means of (the inverse of) an analytic function $\zeta = \zeta(w)$, a harmonic function $u(\xi, \eta) = \Re f(\zeta)$ regular in D is transformed into the harmonic function $\Re f\{\zeta(w)\}$ regular in D′. This simple remark explains the importance of Riemann's Mapping Theorem for the applications of conformal mapping to fluid dynamics and the like. If, for instance, the conformal mapping of D onto D' is known to extend to a continuous mapping of the *closure*[4] of D onto that of D', then this mapping transforms the solution of the first boundary-value problem for the region D into the solution of the corresponding problem for D'. If, in particular, D is a circular disc, then use can be made of the Poisson Integral studied above.

These remarks make it plain that it will be profitable also for potential theory to study, as we shall do in the next section, how the conformal mapping of a given region onto a circular disc behaves on the boundary. We shall need for this purpose a lemma from potential theory, with which we shall conclude the present section.

LEMMA. *Let there be given a harmonic function $u(x, y)$ regular in a region D, and continuous in the extended domain obtained by adding to D the points of an analytic arc that belongs to the boundary of D. Let $u = 0$ on this arc. Then $u(x, y)$ can be continued beyond this analytic arc; this means that every point of the arc has a neighborhood such that there exists a harmonic func-*

[4] The *closure* of a given point-set is obtained by adding to the given set those of its points of accumulation that did not belong to it in the first place.

tion which is regular in the region consisting of D and this neighborhood and which coincides in D with $u(x, y)$.

Proof. If the analytic arc on the boundary of D is represented by $z = z(t)$, $\alpha \leqq t \leqq \beta$, then the function $z(t)$ maps a neighborhood K in the t-plane of every point t_0 in $\alpha \leqq t_0 \leqq \beta$ onto a neighborhood U of the point $z = z(t_0)$; this mapping is conformal, and if we confine ourselves to regular points $z(t_0)$ of the curve, i.e. to points at which $z'(t_0) \neq 0$, then the mapping is also simple.[5] (The points of the curve at which $z'(t) = 0$ are included, along with corners, among the singularities of the boundary.)

By the inverse of the mapping just mentioned, the given harmonic function, which is regular in the intersection (common part) of D and U and continuous in that region plus the boundary arc, is transformed into a harmonic function $\overline{u}$ that is regular in the part of K on one side of the real axis of the t-plane, continuous in that part augmented by its boundary along the real axis, and zero on the real axis. For the sake of simplicity, let us assume that K is a circular disc of radius R and that the t_0 under consideration is the origin of the t-plane as well as the center of K. Let us apply the Poisson Integral to this disc K. To this

[5] Cf. footnote 1 in § 17 for the definition of a *simple* (or *schlicht*) *mapping.*

end, we use the values of $\overline{u}$ as boundary-values on one of the semi-circles, and the values $-u$ at points symmetric with respect to its diameter on the other semi-circle; the boundary values $\overline{u}(\vartheta)$ thus defined for $-\pi \leqq \vartheta \leqq \pi$ satisfy $\overline{u}(-\vartheta) = -\overline{u}(\vartheta)$. Then (5) can be used, setting $t = t_1 + it_2$, to define a harmonic function $\overline{\overline{u}}$ regular in K, as follows:

$$\overline{\overline{u}}(t_1, t_2) = \frac{1}{2\pi} \int_0^{-\pi} \overline{u}(\vartheta) \frac{(R^2 - r^2)\,d\vartheta}{R^2 + r^2 - 2rR\cos(\vartheta - \varphi)}$$

$$+ \frac{1}{2\pi} \int_0^{\pi} \overline{u}(-\vartheta) \frac{(R^2 - r^2)\,d\vartheta}{R^2 + r^2 - 2rR\cos(-\vartheta - \varphi)}.$$

Since $\overline{u}(-\vartheta) = -\overline{u}(\vartheta)$, it follows that for $\varphi = 0$ and for $\varphi = \pi$, i.e. on the real axis of the t-plane, $\overline{\overline{u}}(t_1, 0) = 0$ holds. Hence the harmonic function represented by the above Poisson Integral, being regular in K and continuous on the closure of K, assumes on the periphery of a semi-circular region the same boundary values as does $\overline{u}$, and it therefore coincides with $\overline{u}$ in the interior of the semi-circular region as well. Thus we have in $\overline{\overline{u}}$ the desired "continuation" of $\overline{u}$. Going back to D (by means of the simple conformal mapping described at the beginning of this proof), we arrive at a continuation throughout U, and beyond D, of the given harmonic function $u(x, y)$, and our lemma is proved.

§ 21. The Correspondence Between the Boundaries under Conformal Mapping

H. A. Schwarz was one of the first investigators to occupy himself with the problem of *how* the points of the boundary of a given simply-connected region correspond to the points of the boundary (periphery) of a circle onto which the region is mapped conformally. His principal result is the following:

Let a region D, bounded by a finite number of arcs of analytic curves,[1] *be mapped one-to-one onto the interior of a circle by means of an analytic function. Then there is a one-to-one continuous mapping, which is an extension of the given analytic mapping, of the closure*[2] *of D onto the closure of the circular disc. This mapping is regular, and has a non-vanishing derivative at every point other than a corner, of each analytic boundary arc.*

The method we shall use in proving this theorem is due to Schwarz himself.

Proof. Let the region D lie in the ζ-plane, and let $w = f(\zeta, z)$ be a function that gives a simple mapping of D onto $|w| < 1$ under which $\zeta = z$ goes into $w = 0$. We note first that the Green's

[1] Any such arc is represented by an analytic function $z(t)$ having the property that $z'(t) \neq 0$ along the arc.

[2] Cf. footnote 4 of the preceding section.

Function $\log |f| = -g$ is continuous, except at $\zeta = z$, on the closure of D, provided we set $g = 0$ on the boundary. This follows from the fact that $|f|$ is continuous on this closure provided we set $|f| = 1$ on the boundary, and this in turn follows from the uniform convergence of $|f|$ to unity as ζ approaches the boundary of D. For outside the pre-image of a circle about $w = 0$ with radius $1 - \varepsilon$, the value of $|f|$ obviously lies between $1 - \varepsilon$ and 1.

Hence, by the lemma at the end of the preceding section, we can continue the Green's Function beyond every regular boundary point. Therefore there is an analytic function $f^*(\zeta)$ that is regular in the region D augmented by a suitable neighborhood of each of its regular boundary points, and for which $\log |f^*| = -g$ holds in D. Thus, recalling the Cauchy-Riemann differential equations, the analytic function $\log f^*$ is completely determined to within a pure imaginary additive constant. In D, therefore, f^* agrees with f to within a constant factor of modulus unity. If this factor is chosen such that the derivatives at $\zeta = z$ of f and f^* agree, then $f = f^*$ holds everywhere in D. We have thus obtained an analytic continuation of f beyond the boundary of D, and we have also shown that the mapping it represents is continuous at every regular boundary point.

The derivative of f can not be zero at a regular

boundary point; for since D itself is mapped onto the *interior* of $|w| < 1$, the mapping must be simple in the neighborhood of every regular boundary point.

It remains to investigate the corners of the boundary. We shall show that they too are points of continuity of the mapping, and that to each corner there corresponds a well-determined point on $|w| = 1$.

To see this, let us consider a corner P along with the two analytic arcs of the boundary that meet at P, and let us traverse each of these arcs in the direction *toward* P, so that the region D lies to our left as we traverse the one, and to our right as we traverse the other. The images of these two oriented arcs are two oriented arcs, directed toward each other, on the periphery of the unit circle $|w| = 1$; for along each arc of the boundary of D, the mapping function has a non-zero derivative and the mapping therefore transfers a given direction along the entire arc to its entire image arc. The first thing we must prove is that the two image arcs on $|w| = 1$ meet at some definite point P'. Suppose this were not so; let us draw arcs of circles about P as center, cutting sub-regions out of D, and let their radii go to zero. These circular arcs are mapped onto certain curves in $|w| < 1$ that join the images on $|w| = 1$ of the two given boundary arcs;

and unless the image arcs on $|w| = 1$ meet at a point, these curves must approximate an arc A on $|w| = 1$ as the radii of their pre-images go to zero. This means that the inverse of our mapping function, besides being regular in $|w| < 1$ near every point of A, would have to converge uniformly to a constant, viz. to the coordinate z_0 of P, as the arc A is approached from the interior of the circle $|w| = 1$. Upon subtracting z_0 from the inverse mapping function, we would obtain a function that is regular in $|w| < 1$, continuous in $|w| < 1$ augmented by the arc A, and zero on A. By Schwarz' Reflection Principle, we can continue this function beyond the arc A by reflection. But then the function, being zero along a whole arc A interior to its domain of regularity, must vanish identically, which is impossible since the inverse mapping function can not be identically equal to the constant z_0. This proves that the mapping makes the corner P correspond to a single, well-determined point P' on the periphery $|w| = 1$. Finally, the continuity of the mapping at P follows if we once more consider what the mapping does to the above neighborhoods of P, in D, that we formed by drawing arcs of circles about P as center.

The modern theory of the correspondence of boundaries goes well beyond Schwarz' result that we have just proved. Following preliminary results due to Painlevé and Osgood—the former having treated the case of smooth (i.e. continuously differentiable) boundary arcs—the problem was fully solved by the papers of Carathéodory and E. Study. Their proof underwent considerable simplifications at the hands of various authors, in particular Lindelöf and Koebe. For the details, the reader is referred to Bieberbach, *Lehrbuch der Funktionentheorie*,[3] Vol. 2. Here we only give a brief statement of two principal results: *The mapping is continuous on every continuous arc of the boundary that does not intersect itself. The mapping is analytic on every analytic arc of the boundary.*

We mention further that the mapping is isogonal at a boundary point if the boundary curve has a tangent at this point. The mapping function has a derivative at such a point under the further condition that there exist two circles, tangent to the boundary at the point in question, such that in a neighborhood of the point the boundary curve lies between the two circles.

§ 22. Distortion Theorems for Simple Mappings of the Disc $|z| < 1$

The first result pertaining to the problem to be discussed in this section was obtained by Koebe

[3] Chelsea Publishing Co., New York (1945).

in 1907. It may be stated as follows: *If* $w = f(z)$ *is regular in* $|z| < 1$ *and represents a simple*[1] *mapping of this circular disc, then there exists a number* $m > 0$, *independent of* $f(z)$, *such that all boundary points of the image region are at a distance of at least* $|f'(0)| \cdot m$ *from the point* $f(0)$. Briefly stated, this means that under simple mappings $f(z)$ of $|z| < 1$, with fixed $|f'(0)|$, the boundary points of the image region can not come arbitrarily close to $f(0)$; the circular disc of radius $|f'(0)| \cdot m$ about $f(0)$ as center always belongs in its entirety to the image region.

It would be easy to prove this theorem in a few lines—cf., for instance, Bieberbach's *Lehrbuch der Funktionentheorie*, Vol. 2, p. 83 (Chelsea, 1945). Here, however, we shall find it more profitable to use another method which shows at the same time that $m = 1/4$, and which also leads to a whole series of related theorems.

If $w = z + a_2 z^2 + \dots$ gives a simple mapping of $|z| < 1$, then the image region G_r of every circular disc $|z| \leqq r < 1$ omits (i.e. does not cover) a sub-region G_r', containing the point at infinity, of the w-plane. Hence we have

$$(1) \qquad \iint_{G_r'} w^{\nu} \overline{w}^{\nu} \varrho \, d\varrho \, d\varphi > 0,$$

provided the integral converges, and hence

[1] Cf. footnote 1 in section 17.

$$(2) \qquad \lim_{r\to 1} \iint_{G_r'} w^{\nu}\, \overline{w}^{\nu}\, \varrho\, d\varrho\, d\varphi \geqq 0 .$$

Here we have set $w = \varrho e^{i\varphi}$. The integral will certainly converge if $\nu < -1$.

Let us further set $z = re^{i\vartheta}$. On the boundary of G_r', which is the image of $|z| = r$, we set $\varrho = \varrho(\vartheta)$, $\varphi = \varphi(\vartheta)$. We treat (1) as an iterated integral and carry out the integration with respect to ϱ, obtaining

$$(3) \qquad \iint_{G_r'} w^{\nu}\overline{w}^{\nu}\, \varrho\, d\varrho\, d\varphi = \iint_{G_r'} \varrho^{2\nu+1} d\varrho\, d\varphi$$

$$= -\int_0^{2\pi} d\vartheta\, \frac{d\varphi}{d\vartheta}\, \frac{\varrho^{2\nu+2}}{2\nu+2} > 0 .$$

We have

$$\varphi = \arg w = \mathfrak{J} \log w = \frac{\log w - \log \overline{w}}{2i} .$$

Hence we obtain

$$\frac{d\varphi}{d\vartheta} = \frac{w'\,\overline{w}z + \overline{w}'\,w\overline{z}}{2w\overline{w}} .$$

Therefore (3) can now be re-written as follows:

$$(4) \qquad \int_0^{2\pi} \frac{\varrho^{2\nu}(w'\,\overline{w}z + \overline{w}'\,w\overline{z})}{4(\nu+1)}\, d\vartheta < 0 .$$

This is easily transformed into

$$(5) \qquad \int_0^{2\pi} \left(\overline{w}^{\nu+1} z\, \frac{dw^{\nu+1}}{dz} + w^{\nu+1}\overline{z}\, \frac{d\overline{w}^{\nu+1}}{d\overline{z}} \right) d\vartheta < 0 .$$

But we have

$$w^{\nu+1}=z^{\nu+1}\left(1+(\nu+1)a_2z+\left((\nu+1)a_3+\frac{\nu(\nu+1)}{2}a_2^2\right)z^2+\cdots\right).$$

We substitute this series into (5) and integrate term by term, observing that for any non-zero integer k the relation $\int_0^{2\pi} e^{ki\vartheta}d\vartheta=0$ holds. In this way we obtain

$$1+(\nu+1)(\nu+2)\,|\,a_2\,|^2 r^2 \\ +(\nu+1)(\nu+3)\left|a_3+\frac{\nu}{2}a_2^2\right|^2 r^4+\cdots>0.$$

For $r\to 1$, this yields

$$(6)\quad 1+(\nu+1)(\nu+2)\,|\,a_2\,|^2 \\ +(\nu+1)(\nu+3)\left|a_3+\frac{\nu}{2}a_2^2\right|^2+\cdots\geqq 0.$$

We note that the general term on the left-hand side of (6) is of the following form:

$$(\nu+1)(\nu+k)\,|\,a_k+f(a_2,\ldots,a_{k-1})\,|^2.$$

If in particular we take $\nu=-3/2$, then (6) yields

$$1-\frac{|\,a_2\,|^2}{4}\geqq 0.$$

Hence

$$|\,a_2\,|\leqq 2.$$

We also see from (6) that for each coefficient there exists an upper bound. We summarize our results as follows:

If $w=z+a_2z^2+\ldots$ gives a simple mapping of $|\,z\,|<1$, then the coefficients a_k satisfy the inequalities (6) *for every $\nu<-1$. This implies* *the existence of numbers S_k such that $|a_k| \leqq S_k$ holds for every simple mapping of the unit circle. In particular, we have $|a_2| \leqq 2$.*

The bound 2 in $|a_2| \leqq 2$ can not be improved upon, since it is actually attained in the case of the function

$$\frac{z}{(1-z)^2}.$$

This function gives a simple mapping of $|z| < 1$ onto a region whose complete boundary consists of the part from $-1/4$ to ∞ of the negative real axis. For according to § 12, the function

$$\frac{\left(1-\frac{1}{\zeta}\right)^2}{\frac{1}{\zeta}} = \zeta + \frac{1}{\zeta} - 2$$

maps $|\zeta| > 1$ onto a region whose complete boundary consists of the segment from -4 to 0 of the negative real axis.

The function $z/(1-z)^2$ gives the *only* simple mapping of the unit circle for which $a_2 = 2$. This follows by applying (7) below to $\dfrac{1}{\sqrt{w\left(\frac{1}{z^2}\right)}}$.

The best (i.e., lowest) value for S_k that has been obtained so far[2] is ek; thus for any simple

[2] A thorough account of recent progress in this and related problems will be found in the book by A. C. Schaeffer and D. C. Spencer, *Coefficient Regions for Schlicht Functions*, Am. Math. Soc. Colloq. Publ. Vol. 35 (1950). [*Trans.*]

mapping $w = z + a_2 z^2 + \dots$ of $|z| < 1$, we know that

$$|a_k| < ek$$

holds. It is conjectured that the exact bounds are given by $|a_k| \leqq k$. Since

$$\frac{z}{(1-z)^2} = z + 2z^2 + 3z^3 + \cdots + kz^k + \cdots$$

holds, S_k can certainly not be less than k. For the case that all the a_k are *real*, Dieudonné has proved that $|a_k| \leqq k$. At the end of this section we shall see that $|a_k| \leqq k$ also must hold for simple mappings of $|z| < 1$ whose image region is star-shaped.[3] For arbitrary simple mappings of $|z| < 1$, Löwner has proved that $|a_3| \leqq 3$.

For simple mappings

$$z + \frac{\alpha_1}{z} + \frac{\alpha_2}{z^2} + \cdots$$

of the *exterior* $|z| > 1$ of the unit circle, inequalities similar to (6) can be obtained. Arguments very similar to those used above yield

$$1 + (\nu - 1)|\alpha_1|^2 + (\nu - 2)|\alpha_2|^2 + (\nu - 3)\left|\alpha_3 + \frac{\nu}{2}\alpha_1^2\right|^2 + \cdots \geqq 0$$

---

[3] A region is said to be a *star-shaped* (or simply *star*) region if it contains at least one point such that every straight line through this point meets the region either in one line-segment, or in a half-line, or in a full straight line. [*Trans.*]

for every $\nu > -1$. In particular, for $\nu = 0$ we obtain the so-called *area theorem*:

$$(7)\quad |\alpha_1|^2 + 2|\alpha_2|^2 + 3|\alpha_3|^2 + \cdots + n|\alpha_n|^2 + \cdots \leqq 1.$$

Let us return again to the function

$$w = f(z) = z + a_2 z^2 + \ldots$$

regular and simple in $|z| < 1$. Let $f(z) \neq c$ in $|z| < 1$, i.e. assume that the values assumed by $f(z)$ in $|z| < 1$ do not include the number c. Then the function

$$f_1(z) = \frac{cf(z)}{c - f(z)} = z + \left(a_2 + \frac{1}{c}\right) z^2 + \cdots$$

is likewise regular and simple in $|z| < 1$, so that

$$\left| a_2 + \frac{1}{c} \right| \leqq 2.$$

Hence we have

$$|c| \geqq \tfrac{1}{4}.$$

We have thus proved that *no boundary point of the image region of* $|z| < 1$ *under the mapping* $f(z)$ *is closer to the origin than* $1/4$, which establishes the validity of the result stated at the very beginning of this section. At the same time, the constant whose existence is asserted by that result has been determined exactly. In fact, the lower bound $1/4$ that we have found can not be improved upon, since we can exhibit functions for which it is actually attained. This is the case for the function

$$\frac{z}{(1-z)^2}$$

that we also had occasion to consider earlier. This function assumes the value $-1/4$ for $z=-1$.

We can now take up the so-called *distortion theorem*. Once more, let

$$f(z)=z+a_2z^2+\ldots$$

be regular and simple in $|z|<1$. Then

$$g(\zeta)=\frac{f\left(\dfrac{\zeta+z}{1+\bar{z}\zeta}\right)}{f'(z)(1-z\bar{z})}$$

is regular and simple in $|\zeta|<1$, and has a power series expansion valid in $|\zeta|<1$ that begins as follows:

$$g(\zeta)=\zeta+\beta_2\zeta^2+\cdots.$$

Here, $$\beta_2=\frac{1}{2}\left(\frac{f''(z)(1-z\bar{z})}{f'(z)}-2\bar{z}\right).$$

Therefore $$\left|\frac{f''(z)(1-z\bar{z})}{f'(z)}-2\bar{z}\right|\leqq 4.$$

Hence $$\left|\frac{zf''(z)}{f'(z)}-\frac{2|z|^2}{1-|z|^2}\right|\leqq\frac{4|z|}{1-|z|^2}.$$

Thus $$\frac{2|z|^2-4|z|}{1-|z|^2}\leqq\Re\left(\frac{zf''(z)}{f'(z)}\right)\leqq\frac{4|z|+2|z|^2}{1-|z|^2}.$$

But since $$\Re\left(\frac{zf''(z)}{f'(z)}\right)=|z|\frac{\partial}{\partial|z|}\Re\log f'(z)$$
$$=|z|\frac{\partial}{\partial|z|}\log|f'(z)|,$$

we obtain $$\frac{2|z|-4}{1-|z|^2} \leqq \frac{\partial}{\partial |z|} \log |f'(z)| \leqq \frac{4+2|z|}{1-|z|^2}.$$

Integration now yields

$$\frac{1-|z|}{(1+|z|)^3} \leqq |f'(z)| \leqq \frac{1+|z|}{(1-|z|)^3}.$$

These bounds are attained by—once more—the function $z/(1-z)^2$. The result just obtained is called *Koebe's distortion theorem.* It expresses the fact that the ratio of magnification of the mapping can not change very abruptly as we move around in $|z|<1$. The result may also be re-formulated, as follows. Since $f'(z)$ does not vanish anywhere in $|z|<r<1$, its modulus assumes its maximum as well as its minimum on the periphery of this circle; hence for all $|z|<r$, we have

$$\frac{1-r}{(1+r)^3} \leqq |f'(z)| \leqq \frac{1+r}{(1-r)^3}.$$

Therefore if z_1 and z_2 are any two points in this circle, we have

$$\left(\frac{1-r}{1+r}\right)^4 \leqq \left|\frac{f'(z_1)}{f'(z_2)}\right| \leqq \left(\frac{1+r}{1-r}\right)^4.$$

From

$$|f'(z)| \leqq \frac{1+|z|}{(1-|z|)^3}$$

it follows, by integration along the radius-vector of the point z, that

$$|f(z)| = \left|\int_0^z f'(z)\,dz\right| \leqq \int_0^{|z|} |f'(z)|\,d\,|z| \leqq \frac{|z|}{(1-|z|)^2}.$$

Likewise by integration, we may deduce[4] from

$$|f'(z)| \geqq \frac{1-|z|}{(1+||z|)^3},$$

that

$$|f(z)| \geqq \frac{|z|}{(1+|z|)^2}.$$

We thus have the result that

$$\frac{|z|}{(1+|z|)^2} \leqq |f(z)| \leqq \frac{|z|}{(1-|z|)^2}.$$

Once more, these bounds are attained by $z/(1-z)^2$.

The left-hand one of the last two inequalities contains once more the statement of two pages back about the location of the boundary points of the image region. The right-hand one gives a bound for the rapidity of growth of a function that is regular and simple in $|z|<1$. As the periphery of the unit circle is approached, the modulus of the function can become infinite at most like the second power of the reciprocal of the distance from the periphery.

So far, this section has concerned itself with

---

[4] This follows from the fact that every curve that joins $z=0$ to a point z is mapped onto an image curve whose length is at least

$$\frac{|z|}{(1+|z|)^2}.$$

To show this, we may confine ourselves to curves $z=z(s)$ with a continuous derivative $z'(s)$, where s is arc-length. For the length of the image curve, we then have

$$\int^s |f'(z)|\,|z'(s)|\,ds = \int^s |f'(z)|\,ds \geqq \int^{|z|} |f'(z)|\,d|z| \geqq \frac{|z|}{(1+|z|)^2}.$$

arbitrary simple mappings $w = z + a_2 z^2 + \dots$ of $|z| < 1$. Considerable interest attaches to the study of simple mappings *subject to additional special conditions*, such as the requirement that the image region should be *convex*, or that it should be *star-shaped* with respect to $w = 0$ (i.e. that the image region should meet every straight line through $w = 0$ in one line-segment, or half-line, or full line; cf. footnote 3 of this section). For short, we shall refer to such mappings as being, themselves, "convex," or "star," respectively. The two special problems just mentioned are related to each other; for if $f(z)$ gives a convex mapping of $|z| < 1$, then $zf'(z)$ gives a star mapping, and vice versa. To see this, observe that if $f(z)$ gives a convex mapping, then the direction (angle of inclination) of the tangent to the boundary of the image, viz. the angle $\pi/2 + \arg z + \arg f'(z)$, is a monotonically increasing function of $\arg z$ on $|z| = 1$; hence so is $\arg z + \arg f'(z) = \arg zf'(z)$. But this means that $F(z) = zf'(z)$ is a star mapping of $|z| < 1$. The converse is now obvious. Noting that $\arg f' = \Re(-i \log f')$, we obtain the following necessary and sufficient condition for $f(z)$ to give a convex mapping and hence for $F = zf'$ to give a star mapping: On $|z| = 1$,

$$1 + \Re \frac{zf''}{f'} \geqq 0$$

must hold. Since a harmonic function that is regular in $|z|<1$ and positive on $|z|=1$ must also be positive in $|z|<1$, it follows that the circles $|z| \leqq r < 1$ are likewise mapped onto regions that are convex, or star-shaped, respectively.

For analytic functions whose real part is positive, the problem of bounds for the coefficients has been solved long ago. If the results are applied, *via* the condition just derived, to the present case of convex mappings by functions $f(z) = z + a_2 z^2 + \ldots$, it follows that $|a_n| \leqq 1$ must hold for all n. These bounds are exact, since $w = z + z^2 + z^3 + \ldots = z/(1-z)$ maps the disc $|z|<1$ onto the half-plane $\Re w > -1/2$. The above arguments then show that for star mappings by functions $F(z) = z + A_2 z^2 + \ldots$, the inequalities $|A_n| \leqq n$ must hold. These too are the best possible, since $z/(1-z)^2$ represents a star mapping. Hence for star mappings, the inequalities given by the distortion theorem also are the best possible. In the case of convex mappings, however, the distortion theorem is subject to improvement; for such mappings, arguments similar to the ones used earlier yield

$$\frac{1}{(1+|z|)^2} \leqq |f'(z)| \leqq \frac{1}{(1-|z|)^2},$$
$$\frac{|z|}{1+|z|} \leqq |f(z)| \leqq \frac{|z|}{1-|z|}.$$

§ 23. Distortion Theorems for Simple Mappings of $|z| > 1$

We shall deal in this section with, among other things, simple mappings of the region $|z| > 1$ by means of functions that leave the point at infinity fixed. Such functions have a pole of order unity at $z = \infty$. It is not possible to adapt the theorems of the preceding section to the present case by simply transforming the interior of the unit circle into its exterior. For in that section, we were only concerned with functions that were regular in the unit circle, and which therefore excluded a certain number from among their values, viz. the number $w = \infty$. Here, however, we are not imposing any such condition on the functions that map the exterior of the unit circle. To be sure, any given such function will fail to assume certain values; but we are not selecting from the totality of all our mapping functions any subset characterized by the omission of one and the same value omitted by each of its member functions. Thus if we wished to transfer the results to be obtained in *this* section to mappings of $|z| < 1$, we would have to concern ourselves with simple mappings of the interior of the unit circle by means of *meromorphic* functions, i.e. by means of functions that are regular in $|z| < 1$ *except* for one pole.

We shall base the subsequent developments of this section on the following *distortion theorem for the mapping of rectangles*, due to Rengel, which is also of interest in its own right.

Let there be given a rectangle $\mathfrak{R}$ *in the* ζ*-plane, with sides of lengths* a *and* b, *and a conformal mapping* (simple or non-simple) *of* $\mathfrak{R}$ *by means of a function* $F(\zeta)$ *regular and single-valued in the closed rectangle. Let* β *be a lower bound for the lengths of the image curves of the line-segments in the rectangle that are parallel to the sides* b. *Furthermore, let*

$$I = \iint_{\mathfrak{R}} |F'|^2 \, d\xi \, d\eta$$

be the area of the image region of $\mathfrak{R}$ (where $\zeta = \xi + i\eta$). *Then we have*

$$\frac{a}{b} \leqq \frac{I}{\beta^2}.$$

Here the sign of equality holds only in case $F(\zeta)$ *represents a similarity transformation and the above image curves all are of length* β. *If the sign of equality does not apply, then there is in the rectangle a line-segment parallel to* b *whose image has a length greater than* β, *and a positive number* p *dependent only on this line-segment and satisfying*

$$\frac{a}{b} < \frac{I}{\beta^2} - p.$$

Proof. We may assume that $\mathfrak{R}$ is given by $0 \leqq \xi \leqq a,\ 0 \leqq \eta \leqq b$. Then

$$I = \iint_{\mathfrak{R}} |F'|^2\, d\xi\, d\eta = \int_0^a d\xi \int_0^b |F'|^2\, d\eta\,.$$

By Schwarz' inequality,[1] we have

$$\left(\int_0^b |F'|\, d\eta\right)^2 \leqq \int_0^b d\eta \int_0^b |F'|^2 d\eta.$$

Hence

$$I \geqq \frac{1}{b} \int_0^a d\xi \left(\int_0^b |F'|\, d\eta\right)^2.$$

However, the expression

$$\int_0^b |F'(\xi_1, \eta)|\, d\eta$$

[1] Schwarz' inequality states that

$$\left(\int_{x_1}^{x_2} g(x)\, h(x)\, dx\right)^2 \leq \int_{x_1}^{x_2} (g(x))^2 dx \cdot \int_{x_1}^{x_2} (h(x))^2\, dx\,.$$

We recapitulate its proof: Since for any value of λ we have

$$\int_{x_1}^{x_2} (g(x) + \lambda h(x))^2\, dx = \int_{x_1}^{x_2} (g(x))^2 + 2\lambda \int_{x_1}^{x_2} g(x)\, h(x)\, dx + \lambda^2 \int_{x_1}^{x_2} (h(x))^2\, dx,$$

it follows that

$$\left(\int_{x_1}^{x_2} g(x)\, h(x)\, dx\right)^2 \leq \int_{x_1}^{x_2} (g(x))^2\, dx \cdot \int^{2} (h(x))^2\, dx\,.$$

This is Schwarz' inequality. For the sign of equality to hold, there must be a real value λ_1 such that

$$\int_{x_1}^{x_2} (g(x) + \lambda_1 h(x))^2\, dx = 0$$

holds; hence, assuming that $g(x)$ and $h(x)$ are continuous, we must have for this value λ_1, and for all x in $x_1 \leqq x \leqq x_2$, that

$$g(x) + \lambda_1 h(x) = 0\,.$$

represents the length of the image curve of the line-segment $\xi=\xi_1$, parallel to the sides b, in the rectangle $\mathfrak{R}$. Hence by one of our assumptions,

$$\int_0^b |F'|\,d\eta \geqq \beta.$$

Therefore

$$I \geqq \frac{a}{b}\beta^2.$$

If for any $\xi=\xi_1$ we have

$$\int_0^b |F'(\xi_1,\eta)|\,d\eta > \beta + c, \quad c>0,$$

then because of the continuity of $F'(\xi,\eta)$ there exists a $\delta>0$ such that

$$\int_0^b |F'(\xi,\eta)|\,d\eta > \beta + c \quad \text{in} \quad |\xi-\xi_1|<\delta.$$

Thus we now have

$$I > \frac{a-2\delta}{b}\beta^2 + \frac{2\delta}{b}(\beta+c)^2 > \frac{a}{b}\beta^2 + \frac{4\delta\beta c}{b}.$$

This yields

$$\frac{a}{b} < \frac{I}{\beta^2} - \frac{4\delta c}{b\beta}.$$

Now if

$$\frac{a}{b} = \frac{I}{\beta^2}$$

is to hold, then we must have $\delta=0$, i.e. we must have $\int_0^b |F'|\,d\eta=\beta$ for all ξ, and in addition we must have the sign of equality in Schwarz' inequality. By what we have seen in footnote 1 above, this means that for a certain λ, independent of η, we have

$$\lambda + |F'| = 0.$$

This means that $|F'|$ is independent of η. Together with the fact that

$$\beta = \int_0^b |F'|\, d\eta = b\,|F'|$$

is independent of ξ, this means that $|F'|$ is a constant. Therefore[2] the analytic function F' is a constant, and hence F is an integral linear function, i.e. the mapping represented by F is a similarity transformation, Q.E.D.

It is of importance, for what follows, to know that Rengel's distortion theorem, which we have just proved, *remains valid even if we allow* $F(\zeta)$ *to have singularities on the sides* b *of the rectangle* $\mathfrak{R}$ *and on a finite number of slits parallel to these sides provided that the length of each of these slits is less than* b, so that the totality of these slits does not divide the rectangle into two or more separate parts; *and that* I *now stands for the inner content of the image region of* $\mathfrak{R}$, i.e. for the least upper bound of the double integrals

$$\iint_{\mathfrak{R}'} |F'|^2\, d\xi\, d\eta$$

extended over sub-regions $\mathfrak{R}'$, free of singulari-

---

[2] $$\log|F'| = \tfrac{1}{2}\log F' + \tfrac{1}{2}\log \overline{F'} = \mathfrak{R}\log F'$$
is a constant harmonic function, whence $\log F'$ is a constant.

ties but otherwise arbitrary, of the slit rectangle.

To prove this extension of Rengel's theorem, we consider first the case that there are no slits but that $F(\zeta)$ may fail to be regular on the sides b of the rectangle. In this case, we proceed by fixing on a rectangle $\mathfrak{R}_\varepsilon$: $\varepsilon \leqq \xi \leqq a-\varepsilon, 0 \leqq \eta \leqq b$. For this rectangle, the assumptions of the theorem proved above are valid. Denote its area by I_ε. Then

$$\frac{a-2\varepsilon}{b} \leqq \frac{I_\varepsilon}{\beta^2}.$$

Passing to the limit as $\varepsilon \to 0$, we obtain

$$\frac{a}{b} \leqq \frac{I}{\beta^2}.$$

Now if

$$\int_0^b |F'|\, d\eta > \beta + c, \qquad c > 0$$

holds for some ξ within $\mathfrak{R}$, we choose ε so small that this ξ belongs to $\mathfrak{R}_\varepsilon$. The arguments used in the above proof then apply to $\mathfrak{R}_\varepsilon$ without any change, and we obtain

$$\frac{a-2\varepsilon}{b} < \frac{I_\varepsilon}{\beta^2} - \frac{4\delta c}{b\beta}.$$

Passage to the limit as $\varepsilon \to 0$ yields

$$\frac{a}{b} \leqq \frac{I}{\beta^2} - \frac{4\delta c}{b\beta}.$$

If

$$\frac{a}{b} = \frac{I}{\beta^2}$$

is to hold, we infer as above that $F(\zeta)$ must be an integral linear function.

In the general case, where slits are present, we subdivide the rectangle $\mathfrak{R}$, by means of the straight lines that carry the slits, into a finite number of sub-rectangles

$$\mathfrak{R}_\nu : \xi_\nu < \xi < \xi_{\nu+1}, \quad 0 < \eta < b.$$

To each of these we can apply our above argument and obtain

$$\frac{\xi_{\nu+1} - \xi_\nu}{b} \leqq \frac{I_\nu}{\beta^2},$$

where I_ν denotes the inner content of the image region of $\mathfrak{R}_\nu$. Adding all these inequalities, we obtain

$$\frac{a}{b} \leqq \frac{I}{\beta^2}.$$

The sign of equality can hold here only if it holds in each of the constituent inequalities, i.e. for each of the rectangles $\mathfrak{R}_\nu$. Then $F(\zeta)$ must be linear in each $\mathfrak{R}_\nu$. But since the slits do not decompose the rectangle, $F(\zeta)$ can be continued from $\mathfrak{R}_\nu$ to $\mathfrak{R}_{\nu+1}$ and must therefore be linear (and free of singularities) in all of $\mathfrak{R}$.

As an application, we shall now prove the *Grötzsch-Rengel theorem on circular-slit domains: Consider the interior of the plane that has been cut along a finite number of circular arcs with center at the origin, and let* $\mathfrak{w} = f(\mathfrak{z})$ *give a simple and conformal mapping of this interior. Suppose that the following expansion holds at* $\mathfrak{z}=\infty$:

$$f(\mathfrak{z})=a_1\mathfrak{z}+a_0+\frac{a_{-1}}{\mathfrak{z}}+\cdots,\qquad |a_1|=1.$$

Then it follows that

$$|f'(0)|\leqq 1.$$

The sign of equality applies only in the case of rigid motions.

Proof. We first assume that $f(0)=0$. Let r and R be chosen such that all the slits lie in $r<|\mathfrak{z}|<R$. Let $|\mathfrak{w}|=q(r)\cdot r$ be the largest circle about $\mathfrak{w}=0$ that is contained in the image of $|\mathfrak{z}|=r$, and let $|\mathfrak{w}|=Q(R)\cdot R$ be the smallest circle about $\mathfrak{w}=0$ that contains the image of $|\mathfrak{z}|=R$. Now we cut the circular annulus $r<|\mathfrak{z}|<R$ along a radius, and map the resulting region onto a rectangle $\mathfrak{R}$ by means of $\zeta=\log\mathfrak{z}$. The slits (circular arcs) are thereby mapped onto parts of lines parallel to the imaginary axis. We shall now apply Rengel's theorem of p. 168 to the function $\zeta_1=\log f(e^\zeta)$. We have to set

$$a=\log\frac{R}{r},\qquad b=2\pi,\qquad \beta=2\pi,\qquad I\leqq 2\pi\log\frac{Q(R)R}{q(r)r}.$$

Rengel's theorem then yields

$$\frac{\log\dfrac{R}{r}}{2\pi}\leqq\frac{\log\dfrac{Q(R)R}{q(r)r}}{2\pi},$$

i.e.
$$0 \leqq \log \frac{Q(R)}{q(r)}.$$
Since
$$\lim_{r\to 0} q(r) = |f'(0)|, \quad \lim_{R\to\infty} Q(R) = 1$$
holds, the last inequality yields for $r \to 0$, $R \to \infty$ that
$$0 \leqq \log \left| \frac{1}{f'(0)} \right|,$$
i.e. that
$$|f'(0)| \leqq 1.$$
The sign of equality holds here only if $\zeta_1 = \log f(e^\zeta)$ is a similarity transformation. But if
$$\log f(e^\zeta) = A\zeta + B,$$
then
$$f(e^\zeta) = e^{A\zeta} e^B$$
$$f(\mathfrak{z}) = e^{A \log \mathfrak{z}} \cdot e^B.$$
Here A must be a positive integer, since $f(\mathfrak{z})$ is by assumption regular at $\mathfrak{z} = 0$. Thus we have
$$f(\mathfrak{z}) = \mathfrak{z}^A \cdot e^B, \quad A \text{ a positive integer.}$$
Since $f(\mathfrak{z})$ gives a simple mapping, we must finally have $A = 1$, so that
$$f(\mathfrak{z}) = \mathfrak{z} e^B.$$
Lastly, our assumption that at $\mathfrak{z} = \infty$ an expansion of the form
$$f(\mathfrak{z}) = a_1 \mathfrak{z} + a_0 + \frac{a_{-1}}{\mathfrak{z}} + \cdots, \quad |a_1| = 1$$
is to hold, shows that $|e^B| = 1$. Hence the map-

ping is a rotation; if we now drop the auxiliary assumption made at the beginning of this proof, that $f(0)=0$, we see that $f(\mathfrak{z})$ must be a rigid motion if the sign of equality is to hold, Q.E.D.

We shall next prove the *Grötzsch-Rengel theorem on radial-slit domains. Consider the interior of the plane that has been cut along a finite number of segments on straight lines passing through $\mathfrak{z}=0$, and assume that these slits do not contain the points $\mathfrak{z}=0$ and $\mathfrak{z}=\infty$ either as interior or as boundary points. Let $\mathfrak{w}=f(\mathfrak{z})$ give a simple and conformal mapping of this interior, and assume that at $\mathfrak{z}=\infty$ we have an expansion of the form*

$$f(\mathfrak{z})=a_1\mathfrak{z}+a_0+\frac{a_{-1}}{\mathfrak{z}}+\cdots, \qquad |a_1|=1.$$

Then it follows that

$$|f'(0)|\geqq 1.$$

The sign of equality applies only in the case of rigid motions.

Proof. Once again, we first assume that $f(0)=0$. Let all the slits be contained in $r<|\mathfrak{z}|<R$. Let $|\mathfrak{w}|=q'(r)\cdot r$ be the smallest circle about $\mathfrak{w}=0$ that contains the image of $|\mathfrak{z}|=r$, and let $|\mathfrak{w}|=Q'(R)\cdot R$ be the largest circle about $\mathfrak{w}=0$ that is contained in the image of $|\mathfrak{z}|=R$. Further, let q and Q be defined in the same way as they were in the above proof of

the theorem on circular-slit domains. Once again we cut the annulus $r < |\mathfrak{z}| < R$ along a radius and map the resulting region onto a rectangle $\mathfrak{R}$ by means of the function $\zeta = \log \mathfrak{z}$. We then apply Rengel's theorem to the rectangle $\mathfrak{R}$ and the function $\zeta_1 = \log f(e^\zeta)$, this time setting

$$a = 2\pi,\quad b = \log\frac{R}{r},\quad \beta = \log\frac{Q'(R)R}{q'(r)r},\quad I \leqq 2\pi\log\frac{Q(R)R}{q(r)r}.$$

Rengel's theorem then yields

$$\frac{2\pi}{\log\frac{R}{r}} \leqq \frac{2\pi\log\frac{Q(R)R}{q(r)r}}{\left(\log\frac{Q'(R)R}{q'(r)r}\right)^2}$$

$$1 \leqq \frac{\left(\log\frac{Q(R)}{q(r)} + \log\frac{R}{r}\right)\log\frac{R}{r}}{\log\frac{Q'(R)}{q'(r)} + 2\log\frac{Q'(R)}{q'(r)}\log\frac{R}{r} + \left(\log\frac{R}{r}\right)^2}$$

$$\log\frac{Q'(R)}{q'(r)} + 2\log\frac{Q'(R)}{q'(r)}\log\frac{R}{r} \leqq \log\frac{Q(R)}{q(r)}\log\frac{R}{r}$$

$$\frac{\log\frac{Q'(R)}{q'(r)}}{\log\frac{R}{r}} + 2\log\frac{Q'(R)}{q'(r)} \leqq \log\frac{Q(R)}{q(r)}.$$

Now we have

$$\lim_{r\to 0} q(r) = \lim_{r\to 0} q'(r) = |f'(0)|,\quad \lim_{R\to\infty} Q(R) = \lim_{R\to\infty} Q'(R) = 1.$$

Hence as $r \to 0$ and $R \to \infty$,

$$2\log\left|\frac{1}{f'(0)}\right| \leqq \log\left|\frac{1}{f'(0)}\right|$$

$$\log\left|\frac{1}{f'(0)}\right| \leqq 0$$

$$|f'(0)| \geqq 1.$$

The sign of equality applies only if

$$\zeta_1 = \log f(e^{\zeta})$$

is a similarity transformation. This implies, exactly as in the case of the circular-slit domain, that $\mathfrak{w} = f(\mathfrak{z})$ must then be a rigid motion.

By the Riemann Mapping Theorem, there are simple conformal mappings onto the exterior of the unit circle for every simply-connected region that contains $z = \infty$ and has at least two boundary points. In particular, this must hold for circular-slit domains and for radial-slit domains. It is therefore possible to obtain, from the two theorems just proved, new facts concerning simple mappings of $|z| > 1$. We shall in this way obtain a distortion theorem for simple mappings of $|z| > 1$.

Consider any simple conformal mapping $\mathfrak{w} = \varphi(z)$ of $|z| > 1$ whose expansion at $z = \infty$ is of the following form:

$$\mathfrak{w} = \varphi(z) = a_1 z + a_0 + \frac{a_{-1}}{z} + \cdots, \quad |a_1| = 1.$$

We shall find bounds for $|\varphi'(z_1)|$, where z_1 is an arbitrary point in $|z| > 1$. Let $\varphi(z)$ map $|z| > 1$ onto a region D. We may assume here that

$\varphi(z_1) = 0$, since the bounds we are trying to find are for the *derivative* $\varphi'(z_1)$. We now map the region D either onto a circular-slit domain with one slit or onto a radial-slit domain with one slit, by a simple mapping under which $z = z_1$ goes either into the center of the circular arc or into a point on the extension of the straight slit. The mapping function employed for this mapping, being the inverse of the one that figured in the Grötzsch-Rengel theorems above, will be denoted by

$$\mathfrak{z} = f^{-1}(\mathfrak{w}).$$

The function

$$\mathfrak{z} = f^{-1}(\varphi(z)) = F(z)$$

then maps $|z| > 1$ onto one of the two above-mentioned slit domains in such a way that $z = z_1$ goes either into the center of the circular slit or into a point on the extension of the straight slit. We therefore have

$$|F'(z_1)| = \left|\frac{df^{-1}(\mathfrak{w})}{d\mathfrak{w}}\right|_{\mathfrak{w}=0} \left|\frac{d\varphi(z)}{dz}\right|_{z=z_1}.$$

In what follows we shall write either F_k or F_r instead of F, according to whether the image region is the circular-slit domain or the radial-slit domain, respectively; similarly we shall write either f_k or f_r instead of just f. By the Grötzsch-Rengel theorems, we then have

$$|f_r'(0)| \geqq 1, \quad |f_k'(0)| \leqq 1.$$

Hence we obtain

$$|F_r'(z_1)| \leqq |\varphi'(z_1)| \leqq |F_k'(z_1)|.$$

This is a preliminary formulation of the distortion theorem and will presently be supplemented by an explicit calculation of the bounds, i.e. by explicit formulas for the functions F_r and F_k.

The explicit formulas for these mappings of $|z|>1$ onto slit domains of the kind described above can be obtained without difficulty from the discussion in § 12 of the mapping represented by $z+1/z$. We know that this function maps $|z|>1$ onto the plane slit along the line-segment from -2 to $+2$. At the points $z=-1$ and $z=+1$, angles are doubled by the mapping. The points of the real axis are mapped onto points on the extension of the slit. Now if $|z|>1$ is to be mapped onto a radial-slit domain in such a way that $z_1=|z_1|e^{i\vartheta_1}$ goes into a point on the extension of the slit, we observe immediately that

$$F_r(z) = e^{-i\vartheta_1}z + \frac{e^{i\vartheta_1}}{z}$$

does the trick. We find

$$F_r'(z_1) = e^{-i\vartheta_1}\left(1 - \frac{1}{|z_1|^2}\right).$$

The calculation of $F_k(z)$ is a little more laborious. We start from the observation that the function $z + 1/z$ maps all the circles through $z = \pm 1$ onto (doubly-covered) circular arcs and straight lines in such a way that two mutually perpendicular circles through $z = \pm 1$ yield slits that supplement each other to form a full circle, or a full straight line, in the image plane. Since our mapping of $|z| > 1$ onto a circular-slit domain is to leave the point at infinity fixed, and since it is to send $z = z_1$ into the center of the circular arc which forms the slit, we are led to considering the circle about $z = z_1$ as center that is perpendicular to $|z| = 1$. It is the image of just this circle which forms a full circle together with the image of the unit circle. Accordingly, the two points at which our mapping doubles all the angles must be the points of intersection of the unit circle with the circle just described. The radius r of this circle about $z = z_1$ is given by

$$r^2 = |z_1|^2 - 1.$$

Hence the two mutually perpendicular circles are

$$|z - z_1|^2 = |z_1|^2 - 1, \quad |z|^2 = 1.$$

Their points of intersection lie on the straight line

$$z\bar{z}_1 + \bar{z}z_1 - 2 = 0.$$

If we set

$$z = e^{i\vartheta}, \quad z_1 = |z_1| e^{i\vartheta_1},$$

then for the points of intersection we find

$$|z_1|\cos(\vartheta-\vartheta_1)=1.$$

Hence

$$\cos(\vartheta-\vartheta_1)=\frac{1}{|z_1|},\quad \sin(\vartheta-\vartheta_1)=\pm\sqrt{1-\frac{1}{|z_1|^2}}.$$

Thus the points of intersection are

$$\begin{aligned} e^{i\vartheta} &= e^{i\vartheta_1}\cdot e^{i(\vartheta-\vartheta_1)} \\ &= e^{i\vartheta_1}\left(\frac{1}{|z_1|}\pm i\sqrt{1-\frac{1}{|z_1|^2}}\right). \end{aligned}$$

The mid-point of the segment joining these two points is

$$e^{i\vartheta_1}\frac{1}{|z_1|}.$$

The vector joining them is

$$2ie^{i\vartheta_1}\sqrt{1-\frac{1}{|z_1|^2}}.$$

Therefore

$$\frac{z-\frac{1}{|z_1|}e^{i\vartheta_1}}{ie^{i\vartheta_1}\sqrt{1-\frac{1}{|z_1|^2}}}+\frac{ie^{i\vartheta_1}\sqrt{1-\frac{1}{|z_1|^2}}}{z-\frac{1}{|z_1|}e^{i\vartheta_1}}$$

represents the mapping we require, except that we must still multiply it by $\sqrt{1-\frac{1}{|z_1|^2}}$ to satisfy the requirement that the "scale" of the mapping at infinity, i.e. the derivative at $z = \infty$ of $F_k(z)$, should be unity. We thus obtain

$$F_k(z) = \frac{z - \frac{1}{|z_1|} e^{i\vartheta_1}}{ie^{i\vartheta_1}} + \frac{ie^{i\vartheta_1}\left(1 - \frac{1}{|z_1|^2}\right)}{z - \frac{1}{|z_1|} e^{i\vartheta_1}}.$$

Hence we have

$$\begin{aligned} F_k'(z_1) &= \frac{1}{ie^{i\vartheta_1}} - \frac{ie^{i\vartheta_1}\left(1 - \frac{1}{|z_1|^2}\right)}{\left(z_1 - \frac{1}{|z_1|} e^{i\vartheta_1}\right)^2} \\ &= \frac{1}{ie^{i\vartheta_1}} - \frac{\left(1 - \frac{1}{|z_1|^2}\right) i}{e^{i\vartheta_1}\left(|z_1| - \frac{1}{|z_1|}\right)^2} \\ &= \frac{1}{ie^{i\vartheta_1}}\left(1 + \frac{1 - \frac{1}{|z_1|^2}}{|z_1|^2\left(1 - \frac{1}{|z_1|^2}\right)^2}\right) \\ &= \frac{1}{ie^{i\vartheta_1}}\left(1 + \frac{1}{|z_1|^2\left(1 - \frac{1}{|z_1|^2}\right)}\right) \\ &= \frac{1}{ie^{i\vartheta_1}} \frac{1}{1 - \frac{1}{|z_1|^2}}. \end{aligned}$$

The distortion theorem may now be given its final formulation as follows:

THEOREM. Let $w=\varphi(z)$ *be a simple, conformal mapping of* $|z|>1$. *At* $z=\infty$, *let* $\varphi(z)$ *have an expansion of the form*

$$\varphi(z)=a_1 z+a_0+\frac{a_{-1}}{z}+\cdots, \quad |a_1|=1.$$

Then the following inequalities hold in $|z|>1$:

$$1-\frac{1}{|z|^2}\leqq|\varphi'(z)|\leqq\frac{1}{1-\frac{1}{|z|^2}}.$$

The sign of equality holds on the left, at any given $z=z_1$ *in* $|z|>1$, *only for the functions*

$$AF_r(z)+B, \quad |A|=1,$$

and on the right only for the functions

$$AF_k(z)+B, \quad |A|=1.$$

We see at the same time that these functions, for which the sign of equality holds in the distortion theorem, are the *only* simple mappings with an expansion

$$a_1 z+a_0+\frac{a_{-1}}{z}+\cdots, \quad |a_1|=1$$

at $z=\infty$ that map $|z|>1$ onto a circular-slit domain or onto a radial-slit domain in such a way that $z=z_1$ goes into the center of the circular arc or into a point on the extension of the straight slit, respectively.

§ 24. On the Conformal Mapping of Non-Simple, Simply-Connected Regions Onto a Circular Disc

In the course of our work, we have had occasion several times to consider mappings onto a circular disc of regions that are *not* simple (schlicht). Thus, for example, the neighborhood of a branch-point a of the Riemann surface of $w = \sqrt[n]{z-a}$ is mapped by means of $w = \sqrt[n]{z-a}$ onto a simple region which in turn can be mapped onto a circular disc by the procedure of § 18. The same holds for any sub-region of this Riemann surface that is bounded by only one boundary curve. But it is not true that every region that can be cut out by a single boundary curve from *any* given Riemann surface can be mapped onto a circular disc. In fact, we have met examples of Riemann surfaces on which it is possible to draw closed curves that do not divide the surface into separate regions; and if we consider on such a Riemann surface a sub-region that is bounded by a single boundary curve but contains in its interior a non-separating closed curve as just described—say a non-separating polygon—then we should certainly guess that this sub-region can not be mapped one-to-one and isogonally onto the interior of a circle. For, what should become of the non-separating closed curve under such a mapping?

Even if we were to require the mapping to be merely one-to-one and continuous, we could easily see that we would be just as much out of luck. We shall not follow up this train of thought here.

We shall call a region *of planar character* (*schlichtartig*) if it can be mapped one-to-one and continuously onto a simple (schlicht) region. It is shown in topology—the branch of mathematics that deals with the behavior of geometric objects under one-to-one continuous mappings[1]—that a region is of planar character if it is impossible to find in it a closed polygonal train that does not separate the region into two or more parts. We call a region *simply-connected* if every closed curve in the region can be shrunk, within the region, to a point. In the sequel we shall not be concerned with the most general simply-connected regions of planar character but shall deal only with a few particularly simple ones. Specifically, these will be the ones obtainable in the following way: Let there be given a simply-connected region that can be mapped one-to-one and isogonally onto the interior of a circle; this region may be a simple region, say, or a neighborhood of a branch-point of finite order such as was mentioned at the beginning of this section. Let there also be given a second such region having in common with the

[1] Such mappings are also called *topological mappings*. [*Trans.*]

first one a simply-connected simple region. The point-set consisting of all points of the first and second regions just considered is then itself a region. Now consider a third region that can be mapped onto a circular disc and that has a simply-connected, simple region in common with the above union of the first and second regions. We then obtain a new region by forming the union of our third region with the union of the first and second; and so forth. Now the regions we wish to consider are those that can be constructed by a finite number of steps of the kind just indicated. It can be shown that all such regions can themselves be mapped one-to-one and isogonally onto the interior of a circle, provided only that they have more than one boundary point. A simple proof of this fact, based solely on function-theoretic tools, is due to Carathéodory, and we shall devote this section to its exposition. We shall not consider here regions of a more complicated structure; these can be considered as limiting cases of regions of the kind that we have just described.

Now to prove the possibility of the mappings for the above type of region as claimed, it suffices to solve once and for all the following *fundamental problem* (and then to apply its solution a suitable number of times): Let there be given two regions each of which can be mapped one-to-one and isogonally onto the interior of a circle. Let these two regions have a simple, simply-connected sub-region in common. Then we shall show that the *union* of the two given regions, i.e. the region consisting of all points belonging to either of the two given regions, can itself be mapped one-to-one and isogonally onto the interior of a circle. It is no restriction of generality to assume that the boundaries of any pair of regions under consideration each consist of a finite number of analytic arcs and intersect at angles different from zero. For if necessary, the region that is to be mapped may be considered to be a sub-region of a larger region for which the conditions just mentioned are satisfied;[2] then its image will at any rate be a simple region which may then be treated by the methods of the preceding sections. We proceed to the *proof* of the "fundamental problem":

Let $A + B$ be the first and $B + C$ the second of the two given regions, B being their common sub-region. We illustrate their inter-relation in Figs. 38a and 38b (on the following page.)

The region A is bounded by the two arcs α and γ, while C is bounded by β and δ. The common sub-

[2] To prove this in all necessary detail, we have to make use of a theorem proved in § 21 above, viz. the following: If an analytic function maps a region that is bounded by a finite number of analytic arcs onto another such region, then the function is regular on these analytic arcs as well as in the interior of the region, and can therefore be continued beyond these analytic boundary arcs.

region B and its boundary arcs β and γ are reproduced in both Fig. 38a and Fig. 38b. By our above assumption, the curves β and γ meet at their points M and N of intersection at angles different from zero, and to avoid complications we assume M and N are not corners of the boundary of either of the

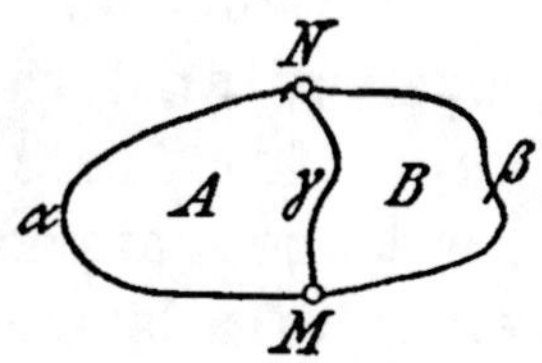

FIG. 38a

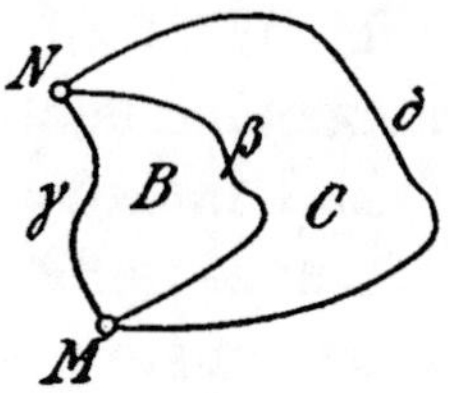

FIG. 38b

given regions. Then if $B + C$ is mapped one-to-one and isogonally onto the interior of a circle, the mapping function must also be regular at M and N. Hence the angles of intersection of the image curve of β with the circular arc that is the image of γ, must be different from zero, being the same as the angles formed by β and γ at their intersections M and N. The region B is mapped onto the sub-region bounded by γ_1 and β_1 of the circular disc (cf. Fig. 39). In this sub-region B_1, we draw a circular arc ε_1 joining M_1 and N_1 and forming with the circular arc γ_1 the angle $\pi/2^n$ (n an integer). In this way, we obtain the convex lens shaded in Fig. 39. This lens has the following property, of importance in the sequel: If we reflect γ_1 in ε_1, we obtain another circular arc ε_2 joining M_1 and N_1 and forming with ε_1 the angle $\pi/2^n$. If we then reflect γ_1 in ε_2, we obtain a circular arc ε_3; the arcs ε_2 and ε_3 bound a lens (or crescent) whose angle of opening is $\pi/2^{n-1}$. Similarly, γ_1 and ε_3 bound a lens (or crescent) of angle $\pi/2^{n-2}$ which when reflected in ε_3 yields another lens (or crescent) of angle $\pi/2^{n-2}$. If we continue this process, we obtain after n steps a covering of the whole circular disc by $n + 1$ such

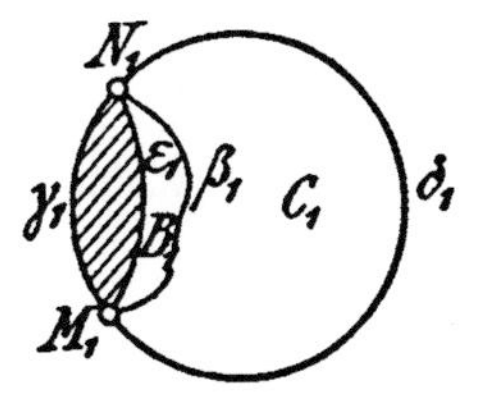

FIG. 39

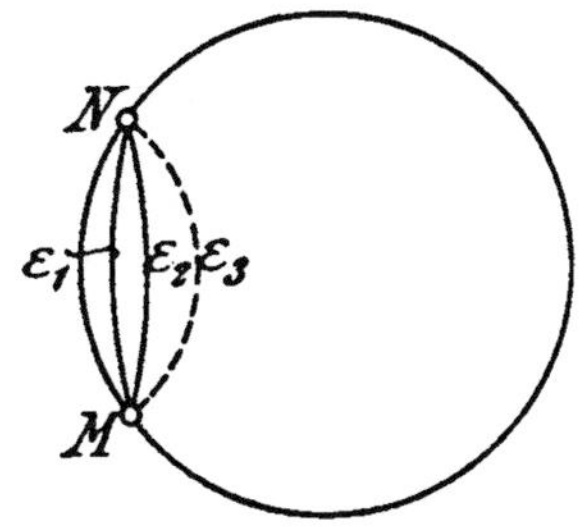

FIG. 40

lenses and crescents (Fig. 40 shows this for the case $n = 3$). To this decomposition there corresponds a decomposition of $B + C$ into $n + 1$ sub-regions $G_1, G_2, \ldots, G_{n+1}$ (Fig. 41 shows this for the case $n = 3$).

Let us now return to the circular arc ε_1 that was the first of those drawn inside the disc. In the region $A + B$, there corresponds to it a curve ε_1' lying in B. Now we map the sub-region of $A + B$

that is bounded by α and ε_1', onto a circle (see Fig. 42). (This is possible, since what we are doing is mapping a simply-connected sub-region of a region $A + B$ which can itself be mapped onto a circle.) The images of α and ε_1' under this mapping will be denoted by α and ε_1. The lens shaded in Fig. 39 is mapped, *via* the region G_1 of Fig. 41, onto the shaded sub-region of the circle in Fig. 42. Now we reflect the interior boundary arc γ_1 of this shaded sub-region in the circular arc ε_1 and

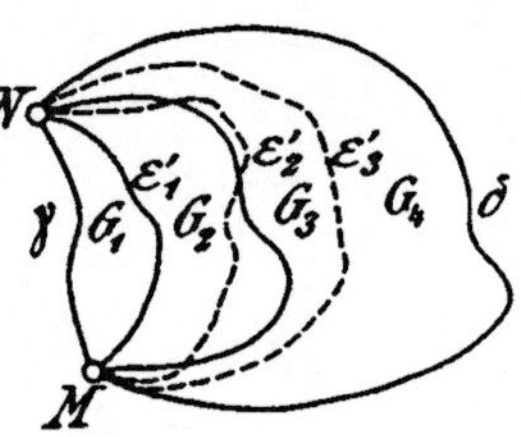

FIG. 41

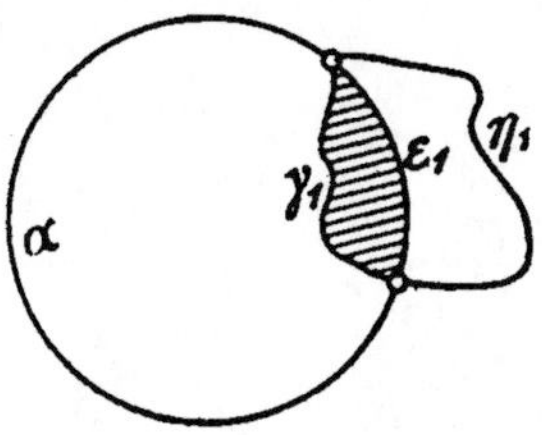

FIG. 42

denote the resulting curve by η_1. In the circle of Fig. 39 onto which we earlier mapped the region $B + C$, there corresponds to the curve η_1 the curve, denoted by ε_2, that is the image of γ_1 under reflection in ε_1. To the lens (or crescent) bounded by ε_1 and ε_2, there corresponds in the region $B + C$ a sub-region that abuts on ε_1. It is this sub-region together with A and G_1, i.e. in all, the region $A + G_1 + G_2$, which under the present mapping (from Fig. 41 to Fig. 42) is mapped onto the simple region bounded by α and η_1. This simple region we now map onto a new circular disc, so that we have, by combination with the preceding mapping, a mapping of all of $A + G_1 + G_2$ onto a circular disc. The image of the curve γ_1 of Fig. 42 is a new curve γ_1' in the new circle; the image of η_1 is an arc η_1' of the periphery of the new circle. If γ_1' is reflected in the latter, we obtain a curve η_2 outside the new circle. The region thus obtained outside the new circle corresponds to the third of the lenses (or crescents) that we constructed earlier in the image circle (Figs. 39 and 40) of $B + C$, and it therefore also corresponds to the sub-region G_3 of $B + C$. The entire new region we now have obtained can in turn be mapped onto yet another circular disc; thus a further part of $B + C$, viz. G_3, has been added to what we can map onto a circular disc. Continuing in this way, we successively include in the mappings onto discs all the regions G_k into which $B + C$ was subdivided above, and after $n + 1$ steps we shall have arrived at a mapping onto a circular disc of the entire region consisting of A together with $B + C$, Q.E.D.

Remark on the Mapping of Non-Simple, Multiply-Connected Regions Onto Simple Regions

For any given such region that is of planar character, the problem of mapping it onto a simple region may be reduced to the case we have just treated. If the given region has n boundary curves, we start by "closing it up" across $n-1$ of these curves (i.e. we start by ignoring that many holes in the region). This yields a simply-connected region that contains the given one in its interior and that can be mapped onto a simple image region by the methods expounded above. Under this mapping, the image of the given multiply-connected region is of course also simple (being a sub-region of a simple region). This image can then be fashioned in various ways into a simply-connected Riemann surface of a finite number of sheets.

§ 25. The Problems of Uniformization

We are now in a position to gather together, and to illuminate from a unifying point of view, the diverse remarks concerning the parametrization of analytic (and in particular of algebraic) curves and functions that we have made in various sections of this book. We shall be concerned especially with pointing up clearly the connection with problems of conformal mapping. We can not, to be sure, include in this introductory exposition complete proofs of the theorems that we shall state; our aim is rather to smooth the path for the reader who wishes to gain an understanding of these deep investigations of Function

Theory. Let us begin by discussing an example:

$z^2 + w^2 = 1$ defines an algebraic curve. We recall two different parametrizations of this circle, viz. $z = \frac{2t}{1+t^2}$, $w = \frac{1-t^2}{1+t^2}$ and $z = \sin\varphi$, $w = \cos\varphi$.

In the first case, consider the Riemann surface of the function $w = \sqrt{1-z^2}$. This surface has two sheets over the z-plane, with branch-points at $z = -1$ and $z = +1$. It can be mapped onto the simple t-plane by, say, the function

$$t = \frac{1+\sqrt{1-z^2}}{z}.$$

But since z and w each assume a definite value at every place on the Riemann surface, there is now associated with every point of the t-plane a definite value of z and a definite value of w. Therefore it must be possible to represent z and w each as a single-valued function of t; and in fact, by solving $t = \frac{1+\sqrt{1-z^2}}{z}$ for z we obtain $z = \frac{2t}{1+t^2}$. Furthermore we have $w = \pm\sqrt{1-z^2} = \pm(zt-1)$. This yields $w = \pm\frac{1-t^2}{1+t^2}$. Thus we have found the above first parametric representation of the circle, except that w as just determined has a double sign. This double sign stems from the fact that the two branches of the function w that differ in sign can be distributed in two ways on the two sheets of the given Riemann surface; the correspondence between the values of w and those of t can thus be set up in two different ways, which is what the double sign expresses. We may therefore, as was done above, simply make a decision as to which one of the signs we wish to use.

Having thus exposed the connection between conformal mapping and the first of the above parametric representations (or, as we shall henceforth say, *uniformizations*) of the circle, we shall find little difficulty in obtaining other uniformizations of this curve. It might, for instance, have seemed even more natural to use the function $t = \sqrt{\frac{z-1}{z+1}}$ for mapping the Riemann surface onto the t-plane. This would have yielded $z = (t^2 + 1)/(1 - t^2)$ and $w = 2it/(1 - t^2)$, a new uniformization of the circle. In this way, we could find quite a few more. It is also quite clear how we can obtain *all* the uniformizations of the circle; we need merely note the following: If one uniformization involves a mapping onto a t-plane, and another uniformization a mapping onto a t_1-plane, then both of these planes are the images of one and the same Riemann surface and must therefore correspond to each other under a one-to-one analytic mapping between the values of t and t_1 that correspond to the same places of the Riemann surface. But such a mapping between two planes, being one-to-one and regular throughout, must be a linear mapping. Therefore *all* possible parameters t can be obtained from *one* of them by means of linear transformations.

So far so good; but can the second of the above parametric representations of the circle be obtained in this way? It can *not*, since it involves trigonometric functions, whereas all the parametric representations to which we have just been led are expressed in terms of rational functions. Wherein lies the difference?

The difference lies in the fact that in all the (rational) uniformizations obtained above, we started from one-to-one mappings of the Riemann surface of $\sqrt{1-z^2}$. But this two-sheeted Riemann surface is by no means the only one on which $w=\sqrt{1-z^2}$ is a single-valued function of the place on the Riemann surface. Let us consider, for instance, the Riemann surface of *four* sheets that has its branch-points at $z=\pm 1$; on this surface too, $\sqrt{1-z^2}$ is single-valued. If we map this surface one-to-one onto a t-plane by means of $t=\sqrt[4]{\dfrac{1-z}{1+z}}$, then we obtain a new parametric representation, one that is not contained in the set we obtained earlier. But neither is it a uniformization by means of trigonometric functions. To arrive at a uniformization of this kind, we start by considering a Riemann surface for which the branch-points $z = \pm 1$ are of order infinity (i.e. a Riemann surface with an infinite number of sheets). The function $\log\left(\frac{z-1}{z+1}\right) = 2ti$ maps this surface onto the simple t-plane. Solving for z and w, we obtain $z = -i \operatorname{cotan} t$, $w = 1/\sin t$. If we take the branch-points to be the two points at infinity of the surface, instead of the points $z = \pm 1$, then $it = \log(iz + \sqrt{1-z^2})$ gives a simple mapping of the Riemann surface of the arcsine function, so that $z = \sin t$, $w = \cos t$.

As this discussion shows, every time we set out to uniformize a given function, we automatically solve the problem of uniformization for a whole collection of other functions. For, every function that is single-valued on the Riemann surface being used, is transformed into a single-valued function of the parameter t under the mapping of the surface onto a simple t-plane, and we have therefore obtained for it a parametric representation. It is clear that these functions are also the only ones to be uniformized by the simple mapping of the Riemann surface, for of course any function that is single-valued in t must have been, before the mapping, a single-valued function on the Riemann surface.

Thus in our first example, the functions that are being uniformized along with the given one are all those that are single-valued and regular everywhere on the Riemann surface of two sheets. All of them become rational functions of t. And since

$$t=\frac{1+\sqrt{1-z^2}}{z}=\frac{1+w}{z}$$

so that t is rational in z and w, we have also given a proof of a theorem from the theory of algebraic functions for this special case.

In our second example, involving the Riemann surface with an infinite number of sheets, the functions that are being uniformized are all those that have branch-points of any orders at $z=+1$ and $z=-1$ and no other branch-points (though they need not be regular at all other points of the plane). All these functions become single-valued functions of the parameter t that was introduced by the mapping of the Riemann surface onto the simple t-plane.

Let us summarize what the two examples show. One alternative is to start with the Riemann surface of two sheets that has its branch-points at $z=1$ and $z=-1$, i.e. with the Riemann surface of an algebraic function defined by $z^2+w^2=1$. We can then solve the following problem: To find a function $t(z)$ such that z, and all other functions which are single-valued and regular on the Riemann surface, e.g. w, become single-valued functions of t. We have also solved the following problems: To uniformize all functions *defined on the surface* that have branch-points of any order at $z = +1$ and $z = -1$, or at $z = \infty$.

The second alternative is to start directly with the simple z-plane, in which case the problems solved above can be formulated, say, in the following way: To uniformize all functions, and only those, that are regular in the plane except at two given points where they may have branch-points of any order. The solution of this problem consists in constructing a Riemann surface on which all the functions in question are single-valued, and in then mapping this surface onto a simple plane. We have also given the solution of a similar further problem, viz. the following: Let there be prescribed a condition on the type of branch-points that the two points ± 1 are to represent; for instance, suppose we want these to be branch-points of order unity, and we then wish to uniformize all corresponding functions (i.e. all functions having branch-points of order unity at $+1$ and -1, and having no other branch-points). We have solved this problem too by constructing the Riemann surface of these functions and then mapping it onto a simple region, which in our case was the full plane.

These formulations immediately suggest generalizations. We may, for example, prescribe branch-points of a different type (i.e. other than those of order unity) at the two points, as in fact we did earlier, in passing. To arrive at problems of a more complicated nature, we must go a little further with our generalizations. Let us, for instance, assign three points of the plane—say, the points 0, 1, and ∞—and then seek, in the set of all functions that have branch-points of any orders at the three given points while being regular everywhere else, a function which is such that all others in the set can be expressed as single-valued functions of the one selected. Does this problem have any solutions, and if so, how can they be found?

The solution leads to the theory of the elliptic modular function that has already been mentioned earlier. Once again, we start by observing that we can without difficulty construct a Riemann surface on which all of the above functions, and no others, are single-valued. To do this conveniently, we first cut the entire z-plane along the real axis and then use one of the two resulting half-planes—say the upper one—for the construction of the Riemann surface. On the boundary of this half-plane, we shall have to deal with three segments, viz. the one from infinity to zero along the negative real axis, then the one from zero to unity, and finally the one from unity to infinity along the positive real axis. We reflect the half-plane in the first of these three segments and join the new half-plane thus obtained to the original one along the same segment. We proceed similarly with the two remaining segments, each time obtaining a further half-plane which is joined to the original one along the segment of reflection. In this way we obtain a region consisting of four half-planes. Each of the three newly created half-planes has two free segments along its boundary, and to these we now attach altogether six further half-planes, by reflection just as before. Then we reflect in the twelve free boundary segments of the new domain, and so forth indefinitely. Now the Riemann surface of infinitely many sheets thus obtained is the one on which all the above functions, and no others, are single-valued. If we succeed in mapping this surface onto a simple region, then the mapping function itself must of course be one of the functions that are single-valued on the surface, and if any other such function is expressed in terms of this mapping function, it must be single-valued also in the new form; hence the problem of uniformization, as posed above, will then be solved, and what we have just done was to reduce this problem of uniformization to a problem of conformal mapping, by means of arguments that belong to *analysis situs* (i.e. topology).

Entirely on the basis of what we have already learned, and without recourse to any new type of argument, we can convince ourselves that the remaining problem in conformal mapping *can* be solved, and what is more, we can indicate in short order how to construct the solution. To begin with, if the Riemann surface is to be mapped onto a simple region, then the image under such a mapping of the original half-plane used in the construction of the surface must also be a simple region. What might this image look like? The isogonality of the mapping breaks down only at the branch-points, whose images will therefore be corners of the image region; since the branch-points are of order infinity, we may expect that the angles at the above corners are zero, just as they are in the case of the mapping of the Riemann surface of the logarithm. The image region of the half-plane under discussion will thus be a triangular region whose three angles are zero. The simplest such triangular region that one might think of is no doubt a triangle formed by arcs of circles. Its three sides must be perpendicular to the circle passing through its three vertices.

Let us therefore try to find the desired mapping by starting with the assumption that the original half-plane does have as its image a triangle of circular arcs as just described. Just which particular such triangle we choose is immaterial, since any two of them can clearly be mapped onto each other by a linear mapping. As a first encouraging fact along the way we recall from § 18 that it is indeed possible to map the circular-arc triangle onto the upper half-plane. By § 21, this mapping is regular also on the boundary, with the exception of the vertices; the latter are mapped onto three points of the real axis. We can of course arrange for these points to be the three points 0, 1, and ∞. Now the transition to one of the further half-planes of the Riemann surface, that is to say, a reflection in a segment of the real axis, corresponds to a reflection in the appropriate side of the circular-arc triangle. If we follow this process of repeated reflections through, we are led to a simple (schlicht) net of circular-arc triangles whose vertices all lie on the circle determined by the vertices of the initial triangle, and whose totality covers the entire circular disc exactly once. Therefore the Riemann surface under consideration has now been mapped one-to-one and isogonally (except at its branch-points) onto the simple interior of a circle. Each individual sheet of the surface, consisting of two half-planes, is mapped onto a quadrangular region bounded by circular arcs, consisting of two circular-arc triangles that are related by reflection. We may think of the whole net as being divided up into such quadrangular regions, each corresponding to one of the sheets of the Riemann surface. The values that z itself takes on are the same in each of the quadrangular regions. Since the passage from any given quadrangular region to any other one always involves an even number of reflections, any two such regions can be mapped onto each other by a linear mapping. All the reflections, and therefore also the linear mappings just mentioned, map our circle onto itself. The totality of these linear mappings constitutes a group that is a subgroup of the elliptic modular group which we mentioned back in § 6. We see immediately that z, when considered as a function of t, is an automorphic function relative to this subgroup. The problem that would lead to the elliptic modular group itself is that of uniformizing all the functions having a branch-point of second order at 0, one of third order at 1, and one of any arbitrary order at ∞.

We shall conclude this section with a few remarks concerning *more general problems of uniformization.* One can either generalize the above problems for the simple (schlicht) plane by adding further branch-points and prescribing all kinds of orders for these, or one can pursue further a different approach, also mentioned above, which consists in passing from the simple plane to the Riemann surfaces of algebraic functions and in treating on these surfaces similar problems as were just outlined for the simple plane itself. We might mention here the problem of *parametric representation of a general algebraic curve.* We have already treated a special case of this, in discussing the elliptic integral of the first kind in § 14 above. We saw in that connection that the elliptic integral of the first kind maps the Riemann surface of the function

$$w = \sqrt{(z-a)(z-b)(z-c)}$$

onto the plane,[2] but that this mapping is not one-to-one, because of the inner structure of the surface (specifically, the occurrence of non-separating closed curves in the interior of the surface) ; z and w turned out to be single-valued automorphic (specifically, doubly-periodic) functions of the integral of the first kind. All functions that have no branch-points on the surface itself were likewise uniformized. It is the Riemann surface common to all these functions that is mapped one-to-one onto the simple plane by the integral of the first kind. If we had found any use for it in § 14, we could easily have constructed this surface from the very beginning right then.

This is in fact the procedure one follows in more general cases. One first constructs the Riemann surface for the class of functions that are to be made single-valued; this class of functions must of course be chosen in such a way that there are no

---

[2] We proved this for the case of real a, b, c; but it remains valid also in the general case. Cf. Bieberbach, *l.c.* Vol. 1.

non-separating closed curves in the interior of its Riemann surface, i.e. that the surface is of planar character, to use a term that we have introduced earlier. For otherwise, the given problem would of course not be solvable. Once we have constructed the Riemann surface, there remains the problem of mapping it onto a simple region. The proof that such a mapping is possible will then complete the proof of the uniformization theorem for the given class of functions.

We may consider, in particular, the problem of *principal uniformization,* i.e. the problem of uniformizing *all* functions that have no branch-points on the Riemann surface of a given algebraic function. As can easily be seen, this problem leads to the construction of a simply-connected Riemann surface that is built up from infinitely many copies of the given algebraic Riemann surface. The new surface can be mapped onto the interior of a simple circular disc; the radius of this disc turns out to be finite if the genus p of the algebraic function is greater than unity, and infinite if $p = 1$. For $p = 0$, it is possible to map the surface onto the whole (extended) plane. As in the above example, the coordinates of the algebraic curve itself then become automorphic functions associated with a certain group. To solve the mapping problem as such, one proceeds by first approximating the surface by an infinite sequence of finitely-sheeted polygons (with straight sides), each contained in the next. We have learned in § 24 how to map such polygons onto a circular disc. It then remains to show that the mapping functions used for this converge to a limit function that maps the entire surface onto a circular disc. In proving this fact, essential use is made of the distortion theorems treated in § 22.

For a further study of uniformization, the reader is referred to Bieberbach, *Lehrbuch der Funktionentheorie,* Vol. 2 (Chelsea 1945); if he prefers to go to the original sources, he is referred to Koebe's papers in the *Mathematische Annalen* and in *Crelle's Journal.* It is Koebe to whom we owe the solution of the problems of uniformization, one of the greatest achievements in the field of Conformal Mapping.

In dealing with the uniformization of arbitrary analytic curves, one soon recognizes the great importance of a problem whose solution for the case of algebraic curves we have indicated above; this is the problem of finding criteria to decide whether the principal uniformization leads to a circle of finite or of infinite radius. The reader will find a comprehensive presentation of this so-called *problem of types* in the book by R. Nevanlinna, *Eindeutige analytische Funktionen* (Berlin 1936; reprint, Michigan, 1944).

§ 26. The Mapping of Multiply-Connected Plane Regions Onto Canonical Regions

The Riemann Mapping Theorem implies that any two simply-connected regions having each at least two boundary points can be mapped onto each other conformally. In view of this fact, it is remarkable that two arbitrary doubly-connected regions can not, in general, be mapped onto each other conformally. Let us consider, for instance, two annuli (each bounded by two concentric circles) that do not have the same ratio of radii. If it were possible to map these two annuli onto each other conformally, then the mapping could be continued by reflection. Making use of the theorem on removable singularities, we see that the extended analytic mapping would be one-to-one in the whole plane, and would therefore have to be linear. Since, furthermore, the point at infinity would stay fixed under this mapping (because of the continuation by reflection that was used), the mapping must be integral linear, i.e. a similarity transformation. But such a mapping leaves the ratio of radii fixed.

However, we shall prove the validity of a different analogue of the Riemann Mapping Theorem: *Every doubly-connected region having no isolated boundary points can be mapped one-to-one and conformally onto a concentric circular annulus.*

Proof. We can make use of the method of § 18. Let us denote the region by D, and let us consider all functions that are regular and simple in D and that map D onto a region whose boundary includes $|z| = 1$ and which has no points in common with $|z| < 1$. Let $\mu(f)$ be the least upper bound of $|f|$ in D, and let ϱ be the greatest lower bound of all the $\mu(f)$. Consider a sequence f_n for which $\mu(f_n) \to \varrho$. We can select from this sequence a uniformly convergent subsequence. Its limit function f gives a mapping of D onto a region D_1 that we shall prove to be simple, and in fact identical with $1 < |z| < \varrho$. That f gives a simple mapping follows from §17, provided only that we know f to be non-constant, which in turn follows from the fact that the amplitude of every approximating function, and therefore also that of the limit function f itself, changes by 2π as a circle containing the unit circle is traversed in D_1. That the image region D_1 is identical with $1 < |z| < \varrho$ is seen as follows. Suppose it were not, and let z_0 be a boundary point of D_1 that is an interior point of $1 < |z| < \varrho$. Making z_0 a branch-point of order two of the exterior of the unit circle, we now map this two-sheeted exterior onto the simple exterior $|z| > 1$ in such a way that one of the two points at infinity remains fixed. Then by Schwarz' Lemma, D_1 is mapped onto a region that is contained in an annulus $1 < |z| < \varrho_1$, with

$\varrho_1 < \varrho$. This contradiction to the definition of ϱ completes the proof.

The result just proved is of fundamental importance also in that it shows the doubly-connected regions to fall into a one-parameter family of classes, according to the value of ϱ, in such a way that any two regions belonging to the same class can be mapped conformally onto each other while regions belonging to different classes can not be so mapped. The conformal invariant $\varrho > 1$ is called the *modulus* of the doubly-connected region.

A similar result also holds for regions of higher connectivities: *Every finitely-connected region having no isolated boundary points can be mapped one-to-one and conformally onto a region bounded by a number of* (complete) *circles.* We shall give a proof of this result at the end of the present section.

There are various other types of canonical regions onto which multiply-connected regions can be mapped; these types include, for instance, regions whose entire boundary consists of certain slits parallel to the real axis, as well as regions bounded by a number of slits that are concentric arcs of circles. Others are slit regions all of whose slits lie on straight lines passing through a fixed point. Finally, some of the slits may be radial ones while the rest are concentric circular arcs. Besides the regions just described, there are a large number of others that can be used as canonical regions. They can frequently be characterized by extremal properties.

Considering the results we obtained in § 23 above, we shall first investigate mappings onto circular-slit domains and radial-slit domains as canonical regions, and we therefore proceed to prove the following theorem:

THEOREM. *Let D be an n-tuply connected region having no isolated boundary points and containing both $z=0$ and $z=\infty$. Then there are just two functions $f_1(z)$ and $f_2(z)$ that give simple mappings of D and satisfy the following additional conditions: $f_1(0)=f_2(0)=0$; at $z=\infty$, each of the functions has an expansion of the form*

$$z+a_0+a_1/z+\ldots;$$

$w=f_1(z)$ maps every component of the boundary onto a circular arc with center at $w=0$ (mapping onto a circular-slit domain); $w=f_2(z)$ maps every component of the boundary onto a straight-line segment whose line passes through $w=0$ (mapping onto a radial-slit domain).

Proof. We consider the set $\mathfrak{M}$ of all functions $f(z)$ that give simple mappings of D, satisfy $f(0)=0$, and also satisfy $f(\infty)=\infty$, $f'(\infty)=1$, i.e. that also have at $z=\infty$ an expansion of the form

$$z+a_0+a_1/z+\cdots.$$

We shall show that there is in this set just one function $f_1(z)$ for which $|f'(0)| = \max.$, and just one function $f_2(z)$ for which $|f'(0)| = \min.$ Then by making use of § 23, we shall see that the function $f_1(z)$ maps D onto a circular-slit domain and $f_2(z)$ maps D onto a radial-slit domain.

To actually carry out the proof along the lines just indicated, we need some preliminary considerations. We must show first that for the functions $f(z)$ of the set $\mathfrak{M}$, the quantity $|f'(0)|$ has a finite upper bound and a non-zero lower bound. For this we use Koebe's distortion theorem (cf. p. 164), as follows. Let $f(z)$ be any function from the set $\mathfrak{M}$, and choose R such that all the boundary components of D lie in $|z| < R$. Then

$$\mathfrak{w} = \frac{R}{f\left(\frac{R}{\mathfrak{z}}\right)}$$

is regular in $|\mathfrak{z}| < 1$ and gives a simple mapping of this disc. By the distortion theorem of p. 164, we therefore know that

$$2/9 \leqq |\mathfrak{w}| \leqq 2$$

must hold on $|\mathfrak{z}| = 1/2$. This shows that on $|z| = 2R$, every function $f(z)$ from $\mathfrak{M}$ satisfies the inequalities

$$R/2 \leqq |f| \leqq 9R/2.$$

This implies, first of all, that the functions $f(z)$

in $\mathfrak{M}$ are bounded in the sub-region of D that is contained in $|z| < R$. Hence by Cauchy's inequalities on the coefficients, the derivatives at $z = 0$ of the $f(z)$ are also bounded. Secondly, we can infer from the above inequalities that there is a non-zero lower bound for $|f'(0)|$, for all f in $\mathfrak{M}$. For if there were not, we could select a sequence of functions in $\mathfrak{M}$ for which $|f'(0)|$ converges to zero. By § 16, we could further select a subsequence that converges uniformly in every interior sub-region of the intersection of D and $|z| < R$. The limit function of this subsequence then has a derivative that vanishes at $z = 0$; and by § 17, the function either gives a simple (schlicht) mapping or is a constant. Its derivative at $z = 0$ being zero, the function must thus be a constant, and since $f(0) = 0$ for all f in $\mathfrak{M}$, this constant must be zero. But this would contradict the fact that on $|z| = 2R$, the moduli of all functions in $\mathfrak{M}$ can not be less than $R/2$.

Now it is easy to complete the proof. Knowing that the functions in $\mathfrak{M}$ are bounded on $|z| = R$, we can start by selecting two sequences in $\mathfrak{M}$ that converge uniformly in $|z| < R$ to two limit functions $f_1(z)$ and $f_2(z)$, respectively, in such a way that $|f_1'(0)|$ equals the least upper bound and $|f_2'(0)|$ equals the greatest lower bound of all $|f'(0)|$ for f in the set $\mathfrak{M}$. Next we can show—

say by using the auxiliary functions $\dfrac{R}{f\left(\dfrac{R}{\mathfrak{z}}\right)}$ already used above—that the two sequences we just selected also converge uniformly in $|z| > R$, and that therefore the two limit functions have at $z = \infty$ expansions of the form

$$z + a_0 + a_1/z + \cdots.$$

By § 17, the limit functions give simple mappings of the region D, and therefore belong to $\mathfrak{M}$.

Now to show that the image of D under $f_1(z)$ is a circular-slit domain, we proceed as follows. Assume that $w = f_1(z)$ did *not* map the component $\mathfrak{B}$ of the boundary of D onto a circular arc with center at $w = 0$; denote the image of $\mathfrak{B}$ by $\mathfrak{B}'$, and denote by D' the simply-connected region, containing the image of D, whose boundary consists entirely of the boundary component $\mathfrak{B}'$ of the image of D. We map D' onto a circular-slit domain by the methods of § 23. The derivative at $z = 0$ of the mapping function exceeds 1 in absolute value, once more by § 23. By combining the last mapping with that given by $f_1(z)$, we would thus obtain a mapping that belongs to $\mathfrak{M}$ and whose derivative at $z = 0$ is larger in modulus than $|f_1'(0)|$. But this would be incompatible with the definition of $f_1(z)$. Hence $f_1(z)$ maps D onto a circular-slit domain. We can prove in the same way that $f_2(z)$ maps D onto a radial-slit domain.

There remains only the uniqueness proof, i.e. the proof that $f_1(z)$ and $f_2(z)$ are the only functions that give mappings of D of the kinds described above. Suppose, say, that D could be mapped onto two different circular-slit domains by functions from $\mathfrak{M}$. Then we consider the mapping of one of these slit domains onto the other; by § 23, the modulus of the derivative at $z=0$ of the mapping function is at most unity. Since this holds for the mapping of the first onto the second as well as for that of the second onto the first of the circular-slit domains, and since the two mappings are inverses of each other, both must have derivatives of modulus 1 at $z=0$. By § 23, the mappings must therefore be rigid motions. Since, further, they leave $z=0$ fixed and have derivative 1 at $z=\infty$, it follows that the two circular-slit domains, as well as the two mapping functions, are identical. The uniqueness proof for radial-slit domains is the same, Q.E.D.

By using the results of § 23, we have at the same time obtained, for regions of finite connectivity, distortion theorems analogous to those for simply-connected regions in § 23. Having shown above that every region of finite connectivity can be mapped onto a circular-slit domain and onto a radial-slit domain, we can repeat the proof given for simply-connected regions in § 23, and we thus obtain the two following theorems:

I. If a given region D contains the points 0 and ∞, then it can be mapped by a function $z + a_0 + a_1/z + \ldots$, "normalized" at $z = \infty$ and leaving fixed $z = 0$ as well as $z = \infty$, onto a region whose boundary consists entirely of slits along lines through $z = 0$ (a radial-slit domain). The mapping function is characterized by having the smallest derivative (in modulus) at $z = 0$, among all functions that give simple mappings of D and that have at $z = \infty$ an expansion as given above.

II. The same region D can be mapped onto a region whose boundary consists entirely of concentric circular arcs with center at $z = 0$ (circular-slit domain), by means of a function $z + a_0 + a_1/z + \ldots$ that leaves both 0 and ∞ fixed. In this case, the mapping function is characterized by having the largest derivative (in modulus) at $z = 0$, among all functions that give simple mappings of D and that have an expansion of the kind indicated above.

We add the following theorem:

III. There is a function $z + a_1/z + \ldots$ that maps the region D onto a parallel-slit domain. The latter may be chosen in such a way, say, that its boundary consists entirely of line-segments parallel to the real axis. The mapping function is characterized by having the largest real part of a_1 among all functions that give simple mappings of D and have expansions of the kind indicated.

Theorem III is proved as follows:

1. Let R be such that $|z| > R$ is contained in D. Then by the area theorem (7) of p. 162, every simple mapping

$$z + \frac{\alpha_1}{z} + \cdots$$

of D satisfies

$$|\alpha_1| \leq R^2.$$

Hence for all simple mappings of the given region D, the real parts of the a_1 are bounded. These real parts therefore have a finite least upper bound.

2. Consider a sequence of simple mappings

$$z + \frac{\alpha_1}{z} + \cdots$$

of D for which the $\Re\, a_1$ converge to the above least upper bound. Using the same arguments as we did in the proofs of the two preceding theorems, we can select a subsequence that converges uniformly to a simple mapping

$$z + \frac{\alpha_1}{z} + \cdots$$

of D for which $\Re\, a_1$ assumes its maximum (in the set of all such simple mappings).

3. Now we shall show that all the components

of the boundary of the image region onto which the above limit function maps D, are slits parallel to the real axis. To this end, we note the following:

a) If we map D by two successive simple mappings of the form

$$z_1 = z + \frac{\alpha_1}{z} + \cdots$$
$$z_2 = z_1 + \frac{\beta_1}{z} + \cdots,$$

then the coefficients of $1/z$ behave in an additive fashion, i.e. we have

$$z_2 = z + \frac{\alpha_1 + \beta_1}{z} + \cdots,$$

as we can verify immediately by calculation.

b) $|z| > R$ is mapped by means of

$$w = z + R^2/z$$

onto a region whose entire boundary consists of the slit $-2R < \Re w < +2R$.

c) As we saw in part 1. of this proof, all simple mappings

$$z + \frac{\alpha_1}{z} + \cdots$$

of D satisfy the inequality $|\alpha_1| \leqq R^2$, hence also $\Re \alpha_1 \leqq R^2$. The sign of equality holds here, by the area theorem, only for the mapping $w = z + R^2/z$.

d) If D is simply-connected, and subjected to simple mappings of the form

$$z+\frac{\alpha_1}{z}+\cdots,$$

then by a) and c) above, $\Re\alpha_1$ assumes its largest possible value in the case that the image region of D has as its entire boundary a slit parallel to the real axis.

e) Now consider the simple mapping of a given multiply-connected region D by means of that function

$$z+\frac{\alpha_1}{z}+\cdots$$

for which $\Re\alpha_1$ assumes its largest possible value, and suppose that the image of a boundary component of D were *not* a straight slit parallel to the real axis. Take the region D_1 whose boundary is the image of just that boundary component; D_1 is simply-connected and contains the image of D. If we map D_1 onto a parallel-slit domain, the mapping function

$$z+\frac{\beta_1}{z}+\cdots$$

satisfies $\Re\beta_1 > 0$, in accordance with d) above. If this mapping is combined with the mapping of D considered just before—i.e. with the one that gives rise to the maximum of $\Re\alpha_1$—then by a), we obtain a mapping of D for which the real part of the coefficient of $1/z$ is even larger than the supposed maximum. This contradiction shows that all the boundary components of the image region of D are straight slits parallel to the real axis, and Theorem III is proved.

We note a *corollary* to what has just been proved: Among all the simple mappings of D of the form

$$z + \frac{\alpha_1}{z} + \cdots ,$$

the one that furnishes a maximum for $\Re \alpha_1 e^{-2i\varphi}$ maps D onto a region whose boundary consists of straight slits in the direction φ. In particular, therefore, we shall have a minimum for $\Re \alpha_1$ if these boundary slits are parallel to the imaginary axis.

Next we shall give a proof for the following theorem, due to Koebe and already stated earlier in this section: *Every region of finite connectivity can be mapped conformally onto a region whose boundary consists of a number of complete circles.*

Proof. We start by mapping the given region onto a circular-slit domain. The latter we reflect in one of its boundary arcs, whereupon we join together the original and its reflection, along the arc in which the reflection took place. Then we reflect the resulting (two-sheeted) region in one of its boundary arcs and join it crosswise, along this arc, to its (two-sheeted) reflection, thus obtaining a region of four sheets. Continuing in this way *ad infinitum*, we build up a covering surface of infinitely many sheets over the original circular-slit domain. This covering surface enjoys the property of being mapped onto itself by any reflection in one of the circular arcs. Now we map this covering surface, by the method described on p. 186ff., one-to-one and conformally onto a simple region. We can do this, for instance, by immediately following up each of the reflections that are used in building up the covering surface, with a simple mapping onto a circular-slit domain, the pre-image in each case being a region of just two sheets. From the resulting sequence of mappings we can select a uniformly convergent subsequence whose limit function gives a simple mapping of the covering surface. The (simple) image region of the covering surface under this mapping has the property of going over into itself under any "reflection" in an image of any one of the circular arcs on the surface; here, "reflection," for the moment, means merely a one-to-one anti-conformal mapping (cf. § 1)—viz., the mapping induced *via* the above simple mapping of the covering surface by one of the (actual) reflections of this surface in one of its circular arcs—but we shall proceed to prove that the images of the circular arcs of the surface all are complete circles, and that therefore the "reflections" of the simple image region are *actual* reflections, in circles; accordingly, we shall then have proved that the given region, which was used as the starting point for our construction of the covering surface, can be mapped onto a canonical region whose entire boundary consists of a number of circles.

Corresponding to the infinitely many sheets of the covering surface, its image region under the mapping described above appears subdivided into infinitely many sub-regions obtainable from each other by the successive "reflections." *We shall show first that the* (infinite) *sum of the squares of the circumferences of these sub-regions is convergent.* To this end, we embed the boundary curves of one of the sub-regions in doubly-connected regions (neighborhoods) and subject these neighborhoods to the "reflections" that carry us into the other sub-regions. Clearly we can choose the neighborhoods so small that they intersect neither each other nor their successive images under the "reflections." Then clearly, the sum of their areas is convergent. Now if we can demonstrate the existence of a number μ independent of the curves to be embedded, and which is such that the circumference U of any of the boundary curves satisfies with the area F of the embedding region (neighborhood) the relation $U^2 < \mu F$, then we may deduce the convergence of the sum of all the U^2 from the convergence of the sum of all the F. That such a number μ

actually exists follows from Koebe's distortion theorem. For since the originally given region is of finite connectivity, all the embedding neighborhoods are obtainable from a finite number of them by conformal or anti-conformal mappings; and since reflections in the real axis change neither areas nor circumferences, it suffices to show the following: There exists a number μ such that for every one of the (finitely many) doubly-connected regions that are to be mapped, and for every simple mapping $f(z)$ of such a region, the area F and the circumference U of its image under $f(z)$ satisfy the inequality $U^2 < \mu F$. Now the distortion theorem[1] implies the existence of two numbers q and Q, independent of the particular region and the particular mapping, that are such that any two points z_1, z_2 of each of the doubly-connected regions to be mapped satisfy the inequalities

$$q < \left| \frac{f'(z_1)}{f'(z_2)} \right| < Q,$$

hence also

$$q < \frac{\operatorname{Max} |f'|}{\operatorname{Min} |f'|} < Q.$$

Now we have

$$U = \int |f'| \, ds < \operatorname{Max} |f'| \, L,$$

[1] The formulation of the distortion theorem as given on p. 164 above applies to mappings of a circular disc. But since a doubly-connected region can always be covered by a finite number of circular discs, the statement in the text follows from repeated applications of the distortion theorem.

where L denotes the length (circumference) of the curve that is being mapped. Furthermore we have

$$F = \iint |f'|^2 dx\,dy > \text{Min}\,|f'|^2 J\,,$$

where J denotes the area of the region to be mapped. It follows that

$$\frac{U^2}{F} < \frac{\text{Max}\,|f'|^2\,L}{\text{Min}\,|f'|^2\,J} < Q^2\frac{L}{J}.$$

Since there are only a finite number of the doubly-connected embedding neighborhoods that are being mapped, there is a μ such that $Q^2\frac{L}{J} \leqq \mu$ holds for all of them. Hence

$$U^2 < \mu F, \qquad \text{Q.E.D.}$$

Let us number the image regions of the sheets of the covering surface in some arbitrary but definite way, and consider a connected sub-region of the over-all image region that consists of the images of certain sheets and includes all the individual image regions whose labels, as just attached above, are below a certain sufficiently large bound. It is clear that the sum of the squares of the circumferences of its boundary curves may be taken to be as small as we please. We shall show that each of the "reflections" is accomplished by a linear function, which will imply that the points that remain fixed under the "reflection" make up the periphery of a circle. To this end, we combine the "reflection" S_1 that we are studying, with a reflection S_2 in the real axis and with a further linear mapping L which takes the image of ∞ under S_2S_1 back to ∞ and which is such that LS_2S_1 has at ∞ an expansion of the form

$$z+\frac{\alpha_1}{z}+\cdots.$$

Under LS_2S_1, the image region of the covering surface goes into a region that is itself made up of sub-regions that are permuted among each other by the "reflections." Therefore for this new region as well, the boundary curves have circumferences the sum of whose squares converges. Now let us consider a sub-region of the image region that excludes only those individual image regions (of individual sheets) whose labels are large. For this sub-region, as well as for its image under $LS_2S_1=f(z)$, the sum of the squares of the circumferences of the boundary curves is as small as we please. We draw a circle K that separates all these boundary curves from ∞, and we apply Cauchy's Integral Formula to the finite region bounded by K and by the boundary curves R_i (whose circumferences have a small sum of squares). We obtain

$$f(z)=\frac{1}{2\pi i}\int_K \frac{f(\mathfrak{z})}{\mathfrak{z}-z}\,d\mathfrak{z}+\Sigma\int_{R_i}\frac{f(\mathfrak{z})}{\mathfrak{z}-z}\,d\mathfrak{z}.$$

Next we show that

$$\Sigma \int_{R_i} \frac{f(\mathfrak{z})}{\mathfrak{z} - z} d\mathfrak{z} = 0$$

holds. On each R_i we set

$$f(\mathfrak{z}) = f(\mathfrak{z}_i) + \varphi_i(\mathfrak{z}),$$

where $\mathfrak{z}_i$ is any point on the boundary curve R_i. Then

$$\int_{R_i} \frac{f(\mathfrak{z})}{\mathfrak{z} - z} d\mathfrak{z} = \int_{R_i} \frac{\varphi_i(\mathfrak{z})}{\mathfrak{z} - z} d\mathfrak{z}.$$

For, we have

$$\int_{R_i} \frac{d\mathfrak{z}}{\mathfrak{z} - z} = 0,$$

since z lies outside R_i. Now

$$\left| \int_{R_i} \frac{\varphi_i(\mathfrak{z})}{\mathfrak{z} - z} dz \right| < \frac{l_i' l_i}{d},$$

where l_i' denotes the circumference of the image of R_i and l_i denotes the circumference of R_i, while d denotes some number larger than the distance of the point z from the curves R_i. (For on R_i, the modulus of the difference between $f(\mathfrak{z})$ and $f(\mathfrak{z}_i)$ is less than l_i', i.e. we have

$$|\varphi_i(\mathfrak{z})| < l_i').$$

Furthermore, we have

$$l_i' l_i \leqq \frac{l_i'^2 + l^2}{2}.$$

Hence

$$\left| \Sigma \int_{R_i} \frac{\varphi_i(\mathfrak{z})}{\mathfrak{z} - z} d\mathfrak{z} \right| < \frac{1}{2d} \Sigma (l_i'^2 + l_i^2),$$

i.e. the left-hand side is arbitrarily small. But since

$$f(\mathfrak{z}) \quad \text{and} \quad \int_K \frac{f(\mathfrak{z})}{\mathfrak{z} - z} d\mathfrak{z}$$

do not depend on the number of the individual image regions (of sheets) of which the region under consideration is made up (i.e. the region to which Cauchy's Integral Formula was applied above), it follows that

$$\Sigma \int_{R_i} \frac{\varphi_i(\mathfrak{z})}{\mathfrak{z} - z} d\mathfrak{z} = 0,$$

and hence that

$$f(z) = \frac{1}{2\pi i} \int_K \frac{f(\mathfrak{z})}{\mathfrak{z} - z} d\mathfrak{z}$$

is regular in K. Outside K, the function $f(z)$ is regular except for a pole of first order at ∞. Therefore $f(z)$ is linear. The linearity of $f(z) = LS_2S_1$ implies that of S_1, and we have proved that S_1 is an actual reflection in a circle, Q.E.D.

The discussion just finished also shows that all mappings of regions of finite connectivity onto canonical regions bounded by a number of circles, are related among each other by linear transformations; for they are related among each other by mappings between two regions each of which admits of an infinite number of reflections.

The papers by Grötzsch in the *Sitz. ber. Sächs. Akad. Wiss.* contain a large number of theorems similar to the above. These papers may be regarded as the most important achievement in the field of conformal mapping of multiply-connected regions.

BIBLIOGRAPHY

1. General Works on the Theory of Functions:

Bieberbach, L., *Lehrbuch der Funktionentheorie* (Vol. I, 4th ed.; Vol. II, 2nd ed.); New York, Chelsea Publishing Co. (1945)

Carathéodory, C., *The Theory of Functions*, 2 Vols.; N. Y., Chelsea Publishing Co. (1953)

Knopp, K., Theory of Functions, 2 Vols.; N. Y., Dover Publications (1950)

2. Specialized Works on Conformal Mapping:

Bergmann, S., *The Kernel Function and Conformal Mapping*; Amer. Math. Soc. (1950)

Carathéodory, C., *Conformal Representation*; Cambridge Univ. Press (1931)

Courant, R., *Dirichlet's Principle and Conformal Mapping* (with an Appendix by M. Schiffer); N. Y., Interscience (1950)

Nehari, Z., *Conformal Mapping*; N. Y., McGraw-Hill (1952)

Schaeffer, A. C., and Spencer, D. C., *Coefficient Regions for Schlicht Functions*, Amer. Math. Soc. (1950)

Study, E., *Vorlesungen über ausgewählte Gegenstände der Geometrie*, Vol. 2: *Konforme Abbildung einfachzusammenhängender Bereiche*. (With W. Blaschke). Leipzig, Teubner (1913)

3. Works on Related Topics:

Ford, L. R., *Automorphic Functions*; N. Y., Chelsea (1951)

Fricke-Klein, *Vorlesungen über die Theorie der automorphen Funktionen*; Leipzig, Teubner (1897-1912)

Klein-Fricke, *Vorlesungen über die Theorie der elliptischen Modulfunktionen*; Leipzig, Teubner (1890/92)

INDEX

ISBN 978-0-8218-2105-3

9 780821 821053

CHEL/90.H